# THE LAST HUMAN SPRING

*Silent Spring II,*
*Origin of Species II,*
*Walden III,*
*Nurturome I*
*Vs. Genome:*
*Breakthroughs!*

## L. S. Heatherly

Published by the Nature-Human Society 2002 & 2009
Copyright © 2002 by L. S. Heatherly

Cover designed by L.S. Heatherly, assisted by Anish V. Adalja.
This book was printed in the United States of America.

Publisher's Cataloging-in-Publication

Heatherly, L. S.
    The last human spring : silent spring II, origin of species II, walden III, nurturome I vs. genome: breakthroughs! / by L. S. Heatherly. – 1ˢᵗ ed.
    p. cm.
    Includes bibliographical references and index.
    LCCN 2002093478
    978-1-4010-6835-6 (HB)
    978-1-4010-6834-9 (PB)
    1-4010-6836-7 (E Book)

    1. Ecology—Philosophy. 2. Human ecology—
Philosophy. 3. Philosophy of nature. 4. Human
evolution. 5. Civilization. 6. Cosmology.
7. Spirituality. 8. Metaphysics. I. Title.

GF21.H43 2002            179'.1
                         QBI01-700478

**To order additional copies of this book, contact:**
Xlibris Corporation      or      Nature-Human Society
1-888-795-4274                   P.O. Box 82366
www.Xlibris.com                  Phoenix, AZ 85071-2366
Orders@Xlibris.com               Nature-humanity@att.net
16017

For the adulterated, polluted, diminished, human soul and spirit—the true origin of the environmental, social, psychological, and spiritual destruction we struggle against.

# PREFACE

FOR THE READER that reads only about one third of any such book, these portions are recommended: Chs. I and II; half of Chs. III, IV, and V; all of Chs. VII and IX; and selections from Chs. X and XI.

A work of such depth and breadth, by one thinker, can, ultimately, never be finished except with the author's death. A second edition of further elaboration, application, and clarification, hopefully, will follow in five to eight year's time. Here, then, by the author's view, is the sixth, complete system of Western philosophy; and the basis for a partially redeeming renaturalization and renurturalization of Western 'civilization'.

The text was originally published Nov. 2001 with a different subtitle. Minor 2002 changes: more proofing, title changes for Chs. III & IX, and increased use of the term *nurturome*.

The Aug. 2009 changes: (a) new photo of a different cliff dwelling on the cover, (b) updated Reviews & Comments Page added preceding the Title Page, (c) more proofing and (d) increased use of the terms, *nurturome* and *human nurturome*. This Author on Humanity and Living Nature stands by all his statements, theories, propositions and theses (written from 1987 to 1999) in this unprecedented, wide and deep classic on the condition of our human species, and our present and future human condition: a book of "absorbing", "enlightening", and "astounding" human literature, bequeathed to the Ages.

# ACKNOWLEDGEMENTS

The author acknowledges Earth's four billion years of life, social mammals, our 300,000 year-old species, Homo sapiens, our 50,000 year-old subspecies, Homo sapiens sapiens, his Earth-folk a'civilization' lineage out of Africa, into Europe, into America's Appalachia; his parents, Martha and Charles E.; relatives, David, Linda, Laura, Chris, Verna, Sandra, Sue, Roger, Dewey.

The author wishes to thank the Phoenix Burton Barr main library, for reference and research assistance; Joan Mertens for years of excellent word processing, and for moral support; friends, Dr. Henry Bock, Margaret V. Arellano, Kenneth E. Pritchett; poets, Karen Odle, Roberta Albinda, George Gilcrease, Elaine Wiggen, Pauline Mounsey, Ken Jones; authors, Henry David Thoreau, R. W. Emerson, L. Mumford, Theodore Roszak, Max Oelschlaeger, George Sessions, Gary Snyder, Paul Shepard, Michael Zimmerman, and Bill Devall (*See also* "Notes.").

As of 2004: acknowledgement of Dr. Richard R Jurin; as of 2009: Darryl K. House, a great reader and friend; his lovely wife, Teri; and Patrick Kurp, Literary Blogger and Reviewer.

# CONTENTS

## Chapter III: THE REVELATION: NURTURECULTURE AND NURTUROME

## Chapter IV: EGOSELF: ORIGINS, DEVELOPMENTS, AND ARISINGS

## Chapter V: NATURAL, AUTHENTIC, HUMAN MIND

## Chapter X: BRIEF APPLICATIONS OF HUMAN AUTHENTICISM

## Chapter XI: THINK VERSE, SOUL LINES, ECOPOETRY

# CHAPTER I

# THE OTHER SPRINGTIME &

# THE THREE BIG BANGS

## A. The Other Springtime

SPRING IS LIVING nature's time of regrowth. Bursting from Earth-stuff (not from unreachable star-stuff), life reasserts purpose, meaning, and beauty in a festival attended by millions of species. But, something is recently amiss. Something is new under our sun: something opposes the Earth-life festival of spring and evolutionary life itself. Within emerged 'civilization', our human spirit, human being, and living nature herself confront a new, alien coldness, an artificial dehumanization and denaturalization. The human spirit's response is *human springtimes.*

The literature and art of 'civilization' are, generally, merely celebrations of our flowers–of the colors of the human spring. They know little of the natural, human, growth cycle of whole, human self and being within living nature. They subjugate our growth, fulfillment, and perpetuation–our spring, summer, and fall–to the flowers of the individual experiencing a personal, quasi springtime– the flowerings of the intellect-egoself. Why this recent, unnatural obsession amidst the human species with art and literature? Life once was the 'art' of being in the flow of living nature! The answer is that we and most of humankind are snowed in for life by the artificial winter of 'civilization'! Flowers symbolize hope and promise–of fruiting human growth and harvesting of human fulfillment. But, we are left clinging to our flowers. 'Civilization' cancels the growth cycle (circle) of life with the line of history and artificial culture. We end up celebrating flowerings of the intellect-egoself rather than the growth, flowering, fruiting, and perpetuation of ecological, whole, human life, being, and self within our world of Earth-life.

Some of the people swept into history (subjugated and op-pressed by city/state) have periodically broken free, back into the warmth and light of a human spring. Most historians and human scientists deny that human spirit and being resurge against 'civili-zation', and, moreover, deny any systemic need for such resur-gence. But, we now know that the wild growth of pure Earth-life has retaken cities in the past through the wild, pure forces and phenomena of living nature and human nature. Moreover, the re-surgence of human nature and spirit is seen microcosmically in the child's resistance to unnatural, alienized culture and nurture; and in the adult's return to relationships with the living land, such as homesteading, gardening, hiking, fishing, hunting, gath-ering (herbs, rocks, seeds, ancient relics, antiques, and other reapprehensions of life and spirit). Macrocosmically, the respringing of human being and spirit is seen whenever 'civilizations' have fallen or declined in size, releasing some city people back into Earth-folk socioculture–back into villages, family farms, herdsmanship, rural craftsmanship, and urban villages retaining and reviving a'civilization', folk, family-community consciousness, folklore, and patterns of Earth-folk culture and lifeway. This perennial resis-tance, micro and macro cosmic, to historical force–resistance from natural childhood, from folk, organic family-community, and from human recoveries of premodern ways and consciousness–has been proven successful in holding substantial, homeground. This is to say, that our pure, natural human being (or beingness) within us, in alliance with the portion of the family or clan that has remained rooted in Earth-folk community's relationship with Earthlifeworld, bides time against 'civilization'.

A clear, unbiased, non-'civilization'centric view of our historic past reveals that the organic paradigm of Earth-folk culture and lifeway has held significant ground against the thrusts of cities, cities that, in general, were destined to be cyclically broken back toward natural communities by living nature's and human nature's counter-surges. Substantial refugee replantings have been typical, though unrecorded by hoaxful history's trail of pen. In our time, urbanites pull off the highway of "progress" onto a road less traveled these days, the back road that still beckons toward what remains of Earth-folk lifeway, and its 5,000 year old counter culture to city culture. They are guided by the rediscovery that *less is more, smaller is better, and more natural is more real.*

Clearly, both the recorded histories of the various 'civilizations', and contemporary, urban societies exhibit cycles of decline and regeneration not in line with living nature's environmental and procreational cycles–and, moreover, not in line with the evolution of humanity and life iteself. Human nature and human beings have come to be subject to a new, major cycle of illregular decay and regeneration; its origin must lie somewhere in 'civilization', or with 'civilization' itself. Urban people, families, communities, societies and entire 'civilizations' experience the decline, decay, and human regrowth of a third cycle alien to living nature, and alien to the nature of her human species. One major episode of a human springtime occurred 3,000 years ago in China. The Taoists, intellectuals leaving the cities, went off into "the wilderness" to learn from living nature and the primitive cultures still existing on mountain ridges. What they learned became the inspiration of Taoism, which was to help quicken certain, other, human springtimes that followed that of the Taoists. Something similar to this birth of Taoism has been happening in the West since the 1960s; though it remains to be seen if a spiritual and philosophical child from it will be carried to term.

Are there human springtimes? Consider just a few indicator movements within the history of 'civilization'. Some 4,000 years ago Abraham led a tribe away from a city into the wilderness. Some 2,500 years ago intellectuals of Taoism left the city for mountain ridges to learn from remaining, primitive tribes. Some 800 to 1,000 years ago the Myans abandoned cities to embrace again primordial village life. Some 100 to 300 years ago America's pioneerism was, in part, the human respringing of Earthly family-community and soul. Indeed, it is hard to avoid a surprising realization from such a non-'civilization'centric review of the historic past. Namely, some of the people have perennially known–through revelation or the soul's instinct–which way to move in the world, which way *to be* in the world. This spiritual-ontological bearing marks no race, no creed, no period of time: it is natural human being and spirit resurging back toward organismic selflifeworld within Earthlifeworld.

Some of us find it refreshing to the spirit when anthropologists tell us (e.g., on PBS) that the Myans–those of them surviving war, famine and disease–about 1,000 years ago, during the span of about 200 years, abandoned, one by one, all but two of their many cities, returning to village life, to a culture and lifeway more natural, ecological, and sustainable than that of the city. A more recent, dramatic episode of a human spring was the West's regeneration, of organic self, family and community, upon breaking back into a more natural lifeworld awaiting them on the more natural continents of America (tragically at the cost of Native Americans and their cultures). There, Western family-community found Earth regions suitable for human roots, free from the weeds and warpings of overgrown culture's accumulated half truths, compromises, misapplications, surrogations, failed experiments, and oppression. This Western episode of spiritual recovery emanating from America is known by many as romanticism, which was surely, one of humankind's major, human springtimes. 'Civilization' fails our human species and hoaxfully disguises the failure with art, science, philosophy and religion achieved from the sacrifice of natural human being and its authentic nature-human harmony, a sacrifice occurring in the form of blood baths, and spiritual, ecological violations of humanity and living nature, all executed by 'civilization'.

*'Civilization' is the simultaneous, ongoing tearing apart and inept reconstruction of living nature's humanity participating within living nature–the destruction of these two inter-functioning spheres. The pathogenesis within us 'civilized' humans spreads farther out into living nature's Earthlifeworld. It dresses over and indoctrinates what it has not been able to devour of life and natural human nature and being. But, it remains only a pretty, arrogant sham set against our primordial, human birthright—of human nature, spitit, and being within living nature.*

The paradigms and patterns of living nature and human nature that, ecologically intergrown, make up her natural, evolved nature-human relationship have been largely abandoned in favor of those provided by science and philosophy using reason of the intellect, alone or primarily, and those provided by religions that additionally explode whole human being–as ecological bioself-family-community-ecoregion–and separate the remnants from evolved Earthlifeworld, which is source of all. The human revelation (presented herein) reveals and describes this spiritual-ontological-dialectical struggle and cycle of human organismicity versus the artificial alienization so distinctive to 'civilization' society in general and to the West in particular. And, the human revelation reveals how and why it is that 'civilization' humanity finds itself experiencing human springs to break through the alien, artificial decays, or 'winters,' of dehumanizing, denaturalizing, and denurturalizing 'civilization'. We discover, then, how and why we have come to our Last Human Spring.

## 1. The Regathering 'Human Tribe'

The idea haunts us from our purest, visionary moments, *We all are one*: stripped of pretense, desires, convention–of all the adulterations from 'civilization'. We are, notwithstanding legitimate, cultural differences, one 'human tribe' or subspecies of primary, authentic humanity. This book joins in a human reawakening that moves multifarious peoples to place our humanness first, to be aware of, respond to, and participate in the healing and recovery from modernity's spreading decay–to join in the quickenings and resurgings of our Last, Human Spring. These people are rejoining the 'human tribe', reconnecting with primary, authentic human being, spirit, and reality; a humanness 50,000 years old by the subspecies measure, and 300,000 years old by our species measure. They experience a reaffirmation and reawareness of a common, root humanness, a human beingness, which is our source and is vital to our preservation and welfare. This human nature, being, spirit and reality precedes our modern, sci-technic, quasi, artificialized, adulterated humanness; just as it precedes the great, cultural diversity of humankind.

That the human spring might long endure amidst the cold, alien, violating, dehumanizing elements of 'civilization'; that we might not be gone with the wind of accelerating sci-technics and commercialism: such is the daring hope we must entertain in this book and in this epoch of the human journey. This volume's purpose, then, is to further the understanding of our authentic, original humanness, the elements and forces destructive to it, and to further some renaturalizations, some renurturings, toward preservation and significant recovery of this humanness.

We, rejoining the 'human tribe', sense the darkest, coldest of ages looming in runaway sci-technics and commercialism, in their alien fantasies, experiments, adventures, curiosities, career ambitions, struggles for power, wealth, domination, illusions of infinite human change and expansion, and deification of rampant, exploding imagination and invention following their divorce from natural, authentic human purpose, spirit, and being.

We, quickening with the Last, Human Spring, turning our human spirit toward the spirit of Earthlifeworld, may see with eyes of spirit–in dream of sleep or trance of dance or song–the soul's dream of a human summertime millenia long. But, how much of humanity will reach that resecured homeground to breath again our pure spirit-breath? We can only wonder; and hope for an unprecedented reconversion and renaturalization of Western consciousness and being.

It is our gift from living nature's evolved, natural human being that our lifebeing-force simply will not give up. From the depth of our being, our spirit stirs with this instinct and intuition: that our ultimate destiny lies sacred within us and our primordial human past, carried around and sometimes through history's domination-subjugation driven empires and alienized cities by our Earth-folk cultures. Our final destiny, as will be painfully discovered, lies within *what we are* and *why we are*.

Indeed, sci-technic, commercial humanity suffers the ever increasing adulterations of sci-technic 'civilization's' artificialization and alienization (beyond mere alienation); nevertheless, our human will and spirit continue to oppose and resist the progressing mental-social-spiritual pollution of our precious humanness. And our human being itself continues an alternating endurance and resistance–sometimes overcoming and transcending–the assault from wild, rampant science and technics, power-wealth driven industry, commercialism, governments, and global, corporate empires; the assault is upon our human spirit, being, heart, soul, and reality of our perennial, natural selflifeworld. To be sure, we have, since the 1960s, uncovered, in conceptual vision, enough human homeground to support a limited optimism.

## 2. The Journey Back Homeward–Partial Redemption

We can't go home again; but, the Last Human Spring within us beholds the rediscovered vision of it, and moves our spirit *homeward*–toward our natural, authentic human being and whole self,

rather than away from it. It has been a long, misundertaken journey for 'civilization'–for the educated minds of it–5,000 years (and
3,000 years for the West). But, we have now come to a time and
place enabling the Western mind (through the deep ecological
paradigm and/or associated world view thinking) to reapproach
the stunning revelation of a forgotten, forsaken home–our natural,
human selflifeworld with our natural, human nature, being, spirit,
identity, culture and reality within Earthlifeworld. We reapproach,
with deep conflict and denial, our humanness (human beingness)
authentic to 300,000 years of Homo sapiens humanity. This book,
presenting the revelation of human authenticism, completes that
journey, bringing alienized, 'civilization' humanity home again (in
revealed awareness) to natural, authentic, human selflifeworld. A
growing number (of laymen and professionals) have moved substantially toward a destiny of recovered humanness; and, some
(partially mentioned below in Ch. IX) are even on the threshold of
the human revelation. They have made key changes in values and
shifts in paradigms and norms (enumerated below). Still, nearly
all lack one or more of a few breakthrough realizations that are
within, and can be beheld through, the human revelation of human authenticism.

Philosophical dualisms, oppositions, conflicts, dichotomies;
these are dissolved within living nature's organismicity and the
natural, human organismicity of human authenticism (the condition). And yet, reason, philosophy, and science (as opposed to
whole intelligence) feed upon these and other illusions. In ultimate reality, in natural, human reality within living nature's reality, there is only one undissolvable, irreconcilable opposition and
conflict: living nature's–Earthlifeworld's–organismicity (about four
billion years old) *versus* the adulterant, artificial, alien penetrations
from 'civilization', particularly Western, which constitute the malig-
nancy-like illness–the pathogenesis–spreading within living nature's
Earthlifeworld, originating in (and particular to) the human
selflifeworld of 'civilization'. Realizing the origins and nature of this
alienism is a vital part of whatever redemption there remains for us.

With sufficiently recovered and renurtured, natural intelligence, spirit and purpose, we move gradually homeward into healing our human being and spirit. May we wish ourselves Godspeed or Earth-spirit–for our Last Human Spring!

## 3. Natural, Human Society

*The natural person, family and community live in an intimate relationship, in a triecologic, organismic, psycho-social-ontological unity, which, in turn, is in intimate unity within Earthlifeworld. People know what each other are feeling and thinking with less need for speaking, because there is more sharing of natural, authentic life and being. 'Civilization' fragments, adulterates and alienizes this unity of ecological, natural human being. 'Civilization' fosters an unprecedented intimacy inward–the egoself with itself and other egoselves and artificial objects and phenomena–while disrupting the other intimacies–with family, community, and Earthlifeworld–essential to the growth and fulfillment of human being and soul. It is from this destruction of human intimacy, love, caring, nurturing and fulfillment that 'civilization' is found to be discredited, illegitimate and transgressive. It is this revelation (as part of the human revelation) about 'civilization's disruption and dysfunction of natural mind, heart, soul, human being, society, living nature, and their relationship that frees our lifebeing-force from its subjugation by historical life-force (which has divorced and separated from our human being-force) to rebond to living nature's lifebeing-force and its evolved, ecological world (field) of living being.*

## 4. Humanity's Primary, Earth-folk Innocence

Dare we profess the people's underlying innocence–our *human, Earth-folk* innocence? Dare we declare the people victims of history's delusion–of manifold tricks and delusions two of which are: (a) the city's siren that leads to the slum or to the stripped-down, middle class neighborhood, trembling and stumbling for our primordial, spiritual-ontological hydes, clothes, the dance steps and rhythm of family-community procession and fulfillment; and (b) alien, historical force that debilitates family-community-land autonomy and perpetuation! Our children are not any longer born free; they are systematically subjugated and warped from birth by alien schemes. Babes in living nature's arms, torn from her breasts are lead into games on the way to some stage of the human circus-zoo. So many artistically decorated choices advertised! But, all our acts are on the stages of the circus-zoo! "Get your act together!" But what kind of choices are these slippery ones offered through-out youth's dietary, mind salad of scientific 'certainty', spiritual-ontological confusion, spiritual frustration, commercial adultera-tion and, and the ego's inflation that eclipses the whole self and being! Is this chaos of exploding, human system, merely stuff for a theory? Or is this the disintegration of humanity, its disenfrachisement from an evolutionary life process from which we separate at the price of human decimation! Have not our per-sonal selves, our families, our communities, our cultures and our particular ancestors emerged from/with our long-running, natural humanness carried through historic time primarily by naturalis-tic, Earth-folk society! Dare we declare that our existence, human being, spirit, and that of our progeny cannot continue for long without sufficiently preserving our root human being, spirit, na-ture and culture resprouting from our Paleolithic being! And that humankind cannot avoid personal and social decline, decay, deci-mation and possibly extinction if generally uprooted from its es-sential humanness!

Thanks to living nature's evolution of human nature, we all have a nature, being, spirit and reality in common, deeper than cultural diversity and deeper than sci-technic, commer-cial manipulation, exploitation, deception and scheme. This human commonality magnificent in us is rooted in the 300,000 years of our species and the 50,000 years of Homo sapiens

sapiens, and substantially rooted in the many Earth-folk nurturecultures coexisting with 5,000 years of 'civilizaton' socioculture.

Our magnificent past is largely hidden from us, buried by history: words written and deeds done by the power-wealth, greedy, lost souls that crushed and buried much of our Paleolithic and Neolithic nurturecultures, and a'civilization', Earth-folk countercultures (counter to citified, alienized, 'civilization' socioculture) along with the perennial freedom carried within them. We of the human kind of being were once born free, and lived free for 1,000, yes even 10,000, generations. We lose our natural, authentic freedom with the emergence of the city–'civilization.' In this artificial construct, the royal, dominating classes gain architecture, art, technology and other material accumulations, and resultingly lose much of their souls and human being (their human freedom to be human beings). Even worse, those subjugated and oppressed by the alien, artificial system and order of the ruling class, thereby, lose their original, natural freedom through false values, idolatry, ideology, mythology, religion and unnaturally, unhumanly conscripted, organized labor–what Louis Mumford called *the human machine*. (For this volume's purposes the term, is strictly speaking, a contradiction.) Nevertheless, peoples' lives were transmuted and devastated through long-term and life-stifling, conscripted, collective labor, technology and commercialism for purposes outside of and destructive to personal-self, family-self, community-self and the nurtureculture of local ecoself-family-community-ecoregion–which is to say, our human selflifeworld natural and authentic to us.

*Every 'civilization' remains a kingdom; no matter the votes cast on the ballot that excludes natural, authentic, human freedom: our nature-human autonomy–whole self, organic family, organic community ecologically intergrown within living nature.* Ultimately, most 'civilized' crimes against the human soul are committed for the sake of the power and wealth of the ruling class (and somewhat less so), for any existing, allied middle class: the same ethos that still rules

the West and its followers, as well as, less brazenly, all 'civilizations'. The crimes against human spirit and being committed by
any 'civilization' are always claimed just (when partially glimpsed
by the educated class) as necessary for the sake of art, concocted
divine will and 'civilization' itself. Even Voltaire, living in the opu-
lence derived from 3,000 years of Western history, was to speak his
deepest confession—and unknowingly a great concession to rural,
Earth-folk socioculture—with the utterance (in his *Essay on the
Morals and the Spirit of Nations*), "The history of the great events of
this world are scarcely more than the history of crimes."

*This book, then, affirms ultimate truth, beauty, goodness, and re-
ality that humanity originally is conscious of as a unity within natural,
ecoself-family-community-ecoregion nurturecultures. It affirms the
relevation that there exists—and substantially within our ability to preserve and gradually recover—natural, authentic, ecologic mind, consciousness, purpose, love, being, spirit, reality, nurtureculture and
selflifeworld within Earthlifeworld; and, that all these are authentic,
nurturing, fitting and belonging to/with our natural, 50,000 year-old
subspecies, Homo sapiens sapiens, and our 300,000 year-old species,
Homo sapiens.*

## 5. Living Nature and the Human Circus-Zoo

Living nature (evolutionary life) and her human nature have
repeatedly brought 'civilization' sociocultures back downward to a
size closer and more in line with her natural, Paleolithic, human
socioculture. She repeatedly effects a significant rectification of the
'civilized' condition, pulling some humans closer again, more in
their place with respect to physical, psychological, social, and spiritual-ontological ecosystems within living nature's embracing, ecological, organismic being. In rough cycles living nature and human nature break 'civilization' downward, back Earthward.
Humanity's artificial and alien enterprise, the city is downsized,
decentralized, and re-ecologized. Some urbanites are re-autonomized

into urban, village communities; others are dispersed back out into family farm, clan community, tribal village or, more recently, small town. Mother nature redirects the humans she can partially rescue back toward what she created us for: to play our evolved-designed, cooperative role in her miracle play of timeless Earthlifeworld (ecosphere-life)–the only play of life we know of and can actually participate in and experience in the universe. The stricken, ruling class flees to another city or accepts its remnant city. Living nature and her human being do have a flinch point; more evidence that living nature, if not an organism strictly speaking, has her own identity, dignity and organismicity. Living nature, at last, flinches, then counter-surges: some of her people are removed back into more simple, natural life, being, and consciousness. Families, clans, villages, tribes, small towns, and urban villages are regenerated and recovered; they are regranted a substantial portion of their lost primordial freedom lying in the ancient order and patterns bearing human nature's seal of approval. We pay in manifold agonies for our alien experiments and self-violations. The greater our alien enterprises–kingdoms, states, empires and sci-technic systems–the greater living nature's and human nature's resultant, corrective and protective surgery on us to retain evolved human beings within her evolving life. Before the city's capture and conversion of us–as well as, after living nature's rectification of us–we appreciate her protective love for us.

*Living nature has offered Earth-folk society as a compromise between delinquent, adulterated, citified culture and pure, Paleolithic, organismic, ecoself-family-community-ecoregion nurtureculture. Living nature–her world (field) of living being–sat down with infected human nature and being; and, they came up with organic, Earth-folk community as the human ground of being to be held against the city's alienism. Earth-folk lifeway and culture together are life's and human being's treatment-therapy for the illness, 'civilization'. If we don't, sincerely and thankfully accept this gesture, this fortress of 'civilization', Earth-folk socioculture, and seize this moment of human evolution at its critical juncture, living nature will eventually have nothing much of us.*

While living nature slowly and patiently reclaims the material buildings, the wood and stone of disproven, abandoned cities, she more quickly reclaims whatever of humanity that can be rehabilitated. She awaites, substantially less penetrated and adulterated, in Earth-folk family, community and selflifeworld to receive refugees from cities besieged by suppressed and oppressed realities of living nature and her human nature. Unable to cure the pathogenesis—'civilization'—within her human nature, living nature's evolved life offers Earth-folk culture and lifeway as a compromise deal, as more natural living, thinking, feeling and consciousness, and more fitting within Earthlifeworld, within living nature's ecosystems.

While a portion of 'civilization' people retain with their alienized selflifeworld a moderate amount of natural, authentic human being, spirit, reality, and selflifeworld; most, from the slums on through the middle and upper classes, endure a general dehumanized, anomalous condition; the pain soothed by medicine, toys, therapies and unabashed hedonism, egotism, and escapism. Regardless of improvements, refinements and new highs for the acts (here and there, now and then), 'civilization' remains a human circus-zoo. The recent movement to naturalize animal zoos, to replace the barred rooms and cages with larger areas imitative of (limited by budget) the natural lifeworld and social environment of animals parallels contemporary actions, as well as, many historical actions, taken to naturalize and humanize the human circus-zoos of 'civilization's' cities. The affluent have the advantage in the movement toward material restorations and renaturalizations toward natural, authentic, human selflifeworld. Alas, the wealthy upper crust, the very breed of quasi people and consciousness that have directed, through the power-wealth ethic and ethos, the human circus-zoo for itself and the middle class: these people especially have the monetary resources to partially renaturalize and rehumanize their fortified hilltops and districts in the expansive, human circus-zoo. They come to reject some basic values and conditions of the circus-zoo while simultaneously embracing and furthering the packaged values and system of the circus-zoo itself;

this in their particular schizophrenia of mind and spirit. On the other hand, they cannot escape the circus-zoo to human freedom: for, they are rooted ideologically, spiritually and emotionally more deeply in the zoo than other urbanites; their addiction is more tart and binding.

Beyond degrees of personal regeneration touched on below, our ultimate escape from the sci-technic, human circus-zoo to human freedom is through our children. Young couples starting families find rural, organic villages, small towns, and organicized, urban villages–communities that can nurture and renurture whole human being in substantial harmony with living nature. They cut most of the major, adulterating, blood vessels coming from sci-technics and commercialism; renounce and partially cleanse the social and cultural pollution already in themselves; protect their children and secure them in nurturing family, community, and ecoregion of ecobeing. We are, thus, partiallly purified through basic purifications of our children. This well done, we can expect that most of our children–about 75% in the Old Order Amish communities of North America–will choose to stay basically within natural, authentic, human freedom rather than be deceived, beguiled and lured into the alien novelty, curiosity, adventure, and intoxication of the falsely shimmering, sci-technic, commercial, human circus-zoo traveling the electronic journey into alienism.

## 6. To Saddle and Bridle Things

The things and phenomena of mechanistic materialism, upon being automatized, achieve their own self-direction and movement. They achieve, with passing generations, an inherited place and developing growth–automatized, sci-technic and commercial development–within the systems of power-wealth socioculture. Ralph Waldo Emerson saw this: "things are in the saddle and ride mankind." This ominous view and prophecy from Emerson over 150 years ago that things (technology, particularly machines) have taken charge of humanity–has been fulfilled further than he could have imagined. But, countless examples found in 5,000 years of history, as well as contemporary "back to the land" and "less is more" movements, testify that the modern idea of progress, the irreversibility of urbanization, and the programmed slogan "You can't stop progress," are propaganda delusions of sci-technic commercialism, its system, its ethos, and its largely unapprehended philosophy of alienism–articulated as a blend of existentialism and futurism. Did not much of the West renaturalize its human beings upon bursting onto the two more natural continents of America (unfortunately at the expense of the more natural Native Americans)! Some in the West are defying the forces of "progress" in a Last Human Spring, and, through this natural, wholesome rebellion, can substantially redomesticate technology, science, and subjugate commercialism to autonomous ecocommunities for human renurturalization and renaturalization toward growth, fulfillment, and perpetuation of human being and soul. Moving in this direction, we recover much of our human sovereignty, purpose and being.

We can buck the machines and power-driven corporate empires off our backs with the regathering of the 'human tribe', or human species, the regathering of human nature and spirit, of human lifebeing-force, and of nurtureculture. Eventually, our spirit and being will triumph–whether, calmly for many of us, or, desperately for a small number of survivors in a last age of diverse, social disintegration and manifold, human decimation. Humankind will saddle up, bridle, take the reigns, confine, subdue and subordinate machines to our humanness, to our living nature-given human nature, purpose, and destiny. Living nature gives these supreme gifts through natural genes and perpetuating, natural nurtureculture and nurturome. (*See* Part 3 of Chapter II and Part 10 of Chapter III). But, we must be moved to *receive* these gifts via our resistance to a hoaxful, illusionary destiny of misperceived, illusional, second 'super humanity' and 'second nature' via artificialization.

Though the democratic revolution and spirit has been largely eclipsed by modernism's sci-technics and commercialism, some of us, through some recovery of our essential humanness and being-force, can once again redefine and substantially recover our lives and consciousness; we can still again revolt, assert and seize our freedom from the coupled dynamics of automatized sci-technics and commercialism (and their electronic media). Some of us have over and over again throughout 5,000 or 10,000 years of 'civilizations', reclaimed much of our freedom from social conventions, conquerers, kings and ruling classes. A small portion of every urban generation sees through power-wealth 'destinies' and moves to regain if necessary, or to hold fast to, much of primary, authentic, human being and reality. We can and must revolt toward freedom one more time with the deepest human revolution: the one against unaccountable, unevaluated, self-incorporating, power-wealth driven, automatized sci-technics and commercialism.

Some such optimism sprouts for those who see that there is more than meets the eye to Post-industrial selflifeworld. Any remaining untransmuted, natural, authentic, human intelligence beholds living nature's creations intertwined or blended with increasing artificial, human creations. It beholds human adaptation to living nature increasingly giving way to human destruction and the displacement of both living nature, and natural, human behavior (personal and social) by artificial objects, behavior and phenomena. For most, alas, it becomes an expedient, easy belief and delusion that the blend of the natural and the artificial is the appropriate, inevitable, historic "extension" representing 'a new natural' order–a "second nature"–of human selflifeworld superior to an older, primary one. Modernization is viewed as collective, human self-improvement. It has become normal, and whatever is normal must be natural, so the reasoning goes. By this logic all the past norms of history were natural including conquesting empires, slavery, monarchy and tyranny. Looking back upon the history of 'civilizations,' one can discern, by transcending modernitycentricism and 'civilization'centricism, that 'civilized' people have a conspicious, citycentric, false view of what living nature is, and of what human nature and being is.

History and anthropology, indeed, disclose that our seemingly 'normal' state of humanity–a blend of naturals and artificials–emerged only about 5,000 years past, increasing incrementally within cities and 'civilizations'. These disciplines, however, stop short of the admission that living nature and her human nature repeatedly break a fair portion of artificialized, alienized society back down *toward* the pre'civilization', prehistoric condition of humanity. But, in fact, living nature has substantially rescued significant numbers of people back into Earth-folk socioculture (the original counterculture to city culture), back into organic families and communities, coexisting with any remaining, less artificialized, smaller cities diminished in their power to dominate, to oppress, to transmute authentic, Earth-folk communities into urban, alienized humanity, *Homo alienus.*

A primary, spiritual-ontological, dialectical struggle, then, has been occurring since the emergence of 'civilization'. It is the struggle of natural human nature and being against emergent, alienized, artificialized, quasi, unhumanized, alienized human nature and being. This entails the struggle of alien, artificial objects and phenomena against natural objects and phenomena. More thoroughly put, this is the struggle of alienized, adulterated, artificialized, quasi, consciousness, behavior, activities, will, purpose, spirit and being against natural non-artificialized counterparts of these. Simply put, it is the alien versus the natural: alienism *versus* human authenticism (natural, authentic human being and spirit). Ultimately, it is alien pathogenesis, transmutation, and decimation of human being *versus* natural, whole human being.

## B. The Three Big Bangs

It is likely the universe started with a Big Bang. But of vaster significance to us, it was followed much later by a Second Bag, and then a Third Big Bang. These later two cosmic explosions bear upon the very meaning, purpose, and beauty of human life and spirit. Only four billion years ago (eleven billion years after the First Big Bang) a Second Big Bang occurred, which dwarfed the First by releasing with the creation of life, in due course, those very qualities our human species has valued for 300,000 years. This eonic explosion of evolving life was destined to create sentient, conscious and self-conscious life forms each with their own reality to be experienced as living beings. Meaningful being, time and place were all created together by the Explosion of Life on Earth. Only in this second sphere, of Earthlifeworld, can beauty, good-ness, love, meaning, and purpose be found and experienced; be-cause it is only here that living beings have flowered forth to expe-rience the only cosmic treasures, the treasures of Earth-life. Life and its treasures grew forth together in the time and place called Earthworld. Life grew richer and formed communities of living beings participating in ecoregions of Earth's organismic ecosphere. But, then there occurred a Third explosion, a Big Bang within the ongoing, magnificent Explosion of Life in Earthworld. This Big Bang was one of destruction. It blew outward from the human mind through selflifeworld and through Earthlifeworld destroy-ing much of life and the treasures of life, including much of hu-man being and spirit. It continues today, in the human realm and outward into Earthlifeworld–and with accelerating destructiveness. It has been imprecisely described variously as dehumanization, exploitation, oppression, evil, injustice, alienation, and ecocrisis.

This third explosion has been increasingly accelerating during the twentieth century. Since this is, if it is not drastically curtailed, of sufficient magnitude to decimate or extinguish the human species and a great part of Earthlifeworld (ecosphere and biosphere), the *why* of this explosion–or *how* does it fit into the generally stable, slow, four billion year growth of evolving Earth-life–cannot be avoided. If we accept that this explosion within living being is the first of its kind in the four billion year history of life, and then acknowledge the staggering effects upon the human species at ground zero and upon the whole of evolved living nature; do we not then have proof beyond an intelligent doubt (reason is a rationalization and defense of 'civilization') that something alien has happened within living nature (the evolution of life) and her human nature! (Chapter IV describes the egoself's role in the emergence of alienism. Chapter V gives a scenario for the origin of alienism in the mind.)

The thesis of alienism, herein, is that something unprecedented has happened to living nature; that she has been penetrated and infected by an alien object or phenomenon, a truly unprecedented event in its unnaturalness to living nature—its anti-lifeness as a destructive process to Earth's lifebeing-force—and in its destructivity to humanity and living nature (ecosphere). The magnitude of the spreading destruction of Earthlife and the debeingization ('dehumanization') of humans is convincing evidence that aliens are here and are destroying a planet they cannot become a part of—cannot fit into its evolved, ecological play of life. This is true, notwithstanding our deep denial stemming from our vested interest in 'civilization', heavy dependence upon it, and addiction to its present, modern form. The aliens have originated by emerging within us. We are the point of origin, and are the emerging aliens to Earthlifeworld.

While the exact scenario or depiction of this Big Bang in the human universe or realm is of mostly academic importance; the revelation that it has occured is crucial. It is crucial that we arrest its ongoing destruction of human selflifeworld and Earthlifeworld. (In making the case for alien objects-phenomena and alienism one may assert their probable existence, or their existence beyond an intelligent doubt. The latter is more in line with this volume's purpose.)

## C. Evolution's Missing Story

That so many people are captivated by the idea of aliens visiting the Earth stems from the fact that aliens are, indeed, here amongst us; more precisely, alien being has penetrated and is grow-ing within our, human selflifeworld. The emergent and growing idea of aliens on Earth reflects the emergence and explosive growth of alien objects and phenomena within our selflifeworld transmut-ing us into beings alien to our home, to living nature's evolution-ary Earthlifeworld; within which we were evolutionally created to dwell. The captivating idea that extraterrestrials are here on Earth bears some truth. But it misplaces the origin of the aliens as com-ing from outer space: when, in reality, they are emerging within us. The truth is not out there; it is within us, and here within our selflifeworld. How the spiritual-ontological, phenomenal equiva-lent to 'DNA' of alien being formed within human being is a fascinating puzzle and challenge that we confront. We have come to be living the high, sci-technic version of *Dr. Jekyl and Mr. Hyde.* But, we are intent on denying our addiction to far out, anti-life, anomalous, monstrous science and technology right up to our breakdown apocalypse (of the person and the 'civilization'). *Our increasing capture by the belief in aliens from outer space represents our denial that alien being is emerging within us. We are the aliens ap-pearing on Earth from inner space, from within our mind and being.*

The evolution of life has been one of life-force and being-force moving together. Evolution requires that each of the vari-ous living beings instinctively live to defend and perpetuate their distinctive species being. The will to live and the will to be one's natural being are inseparable in the evolution of life. Life-force and being-force are intergrown as the evolution of living being. Thus, life is the evolution of various, distinct spe-cies to fit into the evolution of the ecological systems of living nature's life. The living nature of humanity is bequeathed har-mony with living nature until we rebel. 'Civilization' is the great exodus of humanity from living nature and her evolution of life. We have discovered upon an immense revelation: evolution's missing story.

This digression of a 300,000 year-old species out of evolving life; how so? An unprecedented explosion (exponential increase) of anomalous behavior in general–objects-phenomena or aspects of selflifeworld–in a prominent species (such as Homo sapiens sapiens) including anomalous, alienation, is one likely origin of alienism. A species can accommodate or carry only so much anomaly (in number and kind) beyond which it declines and/or mutates. If the mutation or decline is blocked by an unprecedented (in the species, genus and family), key, sociocultural element or phenomenon (not a gene), the life-force conceivably could in response–for the survival of individual organisms (personal lives)–create an alien element-phenomenon carried thereafter within the species', *procreational* (generational) *life,* but remaining outside the *evolutional life* of the species *being* and living nature's world (field) of being. The will to live (life-force) would have separated (overridden) from the will to be one's being (being-force). The life-force would have clashed with and violated being-force, creating unprecedented, alienized being–natural being infected with an alien, socio-psycho-spiritual-ontological element-phenomenon (and subsequently, increasing numbers of these) not fitting into living nature's eco-spiritual-ontological play of life. Since there is no being but living being (–rocks and inanimate planets, merely exist–can't experience being–can't live–), the misfit one is of living being, wherever one wishes to place the start of living being in the evolution of life. The species' life could continue at the price of harboring an alien feature–a behavioral phenomenon–and increasing numbers of these (alien to the species, genus and family of being). Although outside Earth's world (field) of living being (evolving, living being), this species life would continue via life-force, but would carry increasing, alien pathogenesis (analogous to virus, infection or malignancy in organisms) within the species. Moreover, since Earth's natural beings are inherently, (evolved within) interrelational, ecological systems, the infected species would not be able to play correctly the part it had played when healthy–when fully natural. The result would be destructive to other species (and lifeworld phenomena); and destructive to the infected species through diminished ecorelationships (human and nonhuman) that grow, fulfill, and perpetuate the manifestation of whole human being and self in living nature's ecological system(s). The play of life would be alienized, transmuted, and diminished–not merely changed– via the effects from relating with an incompetent, disruptive, non-fitting, alienized species.

It is the view, in this volume, that human transmutation via this scenario commenced five (or ten) thousand years ago in the conspiciously prominent species, Homo sapiens sapiens, diverting its prominence onto a path of destructive dominance of living nature and her human nature.

Any proposed alienism, additionally, calls for a suspicious look at alienation, which occurs within living nature and human nature. Alienation *per se* is not alien because we can observe instances of it in the selflifeworld of primates and mammals where an individual becomes alienated from his family group temporarily or permanently (and when permanently, leading to illness or death). Both instances are observed incidentally in a variety of living nature's species, and surely occurred in (and occurs in remnant) primordial humanity. We can safely assume that this alienation is many millions of years old, a part of living nature's nature. But, suppose for the sake of argument, firstly, it can be established that alienation increasingly occurs *outwardly* through the socio-psycho-spiritual, human selflifeworld of complex, urban 'civilization', eventually becoming exponentially increased in number, duration, and kind of alienation. The quantitative increase would be so great as to be a qualitative change in the nature of the alienation. In fact, in the city, people become alienated–separated spiritual-ontologically– not just within the primary, social units of family and community, but also in secondary relationships (within education, occupation, other institutions, social policy and practices, cultural phenomena, governments, etc.). Suppose, secondly, that, at the same time, alienation increasingly occurrs *inwardly*, fragmenting the bio-self, the elements that constitute the mind-body organism; as we respond by artificially, insufficiently reassembling the debris from these two accellerations of alienization with applications of sociology, psychology, psychiatry, other sciences, and the manifold transmutations of fragmented, transmuted selflifeworld objects and phenomena. This volume of philosophical thought asserts, herein, that these two suppositions are, in fact, occurring. With alienation accelerating both outwardly and inwardly, the result is the transmutation of *alienation* into *alienism* within human selflifeworld: one origin of *alienism* is the explosion of *alienation*, transmuting it into *alienism* of humanity.

In order to deny that this alienism of the city is not alienism, one would be pressed to find it somewhere in four billion years of living nature–somewhere outside the city (or before settled, agricultural communities). Or second, one would have to suppose that the explosions of human selflifeworld and alienation occurring simultaneously are explosions occurring within living nature, that is, explosions of a portion (subrealm) of living nature's realm. To say this presses one back to explain why living nature's evolution of life waited until recently to manifest this explosion, and to what purpose (if one requires one). It is not credible that our ecocrisis and the unprecedented explosion of alien phenomena within our evolved human species are merely coincidental.

# CHAPTER II

# INTRODUCTORY LIGHT

## A. Revelation as Rediscovery

### 1. Sociocultural Revelation

MULTIFARIOUS PEOPLE WITHIN diverse cultures of sci-technic, commercialism (technocracies) come to the realization that modern 'civilization' ("progress") leads the human soul away from its natural, authentic home. They feel increasingly lost and alienated, and are ripe for quickenings. Many have come to share a renewed, soulful purpose and vision. A movement toward human unity has commenced, in our age, that concievably could preserve humankind. As suggested earlier, this movement is nothing less than a fundamental regathering of primary, human spirit and being, a regathering of the 'human tribe'–of our 50,000 year-old Homo sapiens sapiens

Many people experience glimpses, some even partial revelations of their lost, human home—of an underlying selflifeworld, natural and authentic to all humankind. Many search for and discover pathways leading homeward; and approaching home after being long gone, they must keep swallowing their fears, fighting obstacles, and must remember and embrace in the heart what home means–rooted, fulfilling, perpetuating life and being–and keep moving homeward. The last, greatest, human struggle still holds human, tribal ground, still holds our human soul and spirit, despite new and increasing alien, artificial adulterations of our humanness (human beingness).

We can take heart for several reasons. There have been occuring during the past four decades several key shifts in Western values and consciousness that are in line with, or within, the root paradigm shift—ultimately the reality shift and recovery—advocated by the philosophy of this volume. Concisely put, it views human being, spirit, and reality as authentically consisting only of natural, human organics constituting human organismicity within living nature's organismicity. The following shifts in paradigms and values are occurring, some more than others and all too slowly: (a) from the mechanistic world view to the organic world view, (b) from the anthropocentric to the biocentric and ecologic, (c) from the (scientific) study of separated, analyzed parts to the instinctive and intuitive (holistic) understanding of wholes and systems, (d) from the self as an atomistic, independent ego entity (subject) to self as interrelational and interidentifying with the essential objects, people, social units and living nature's systems (from subject-object dualism to subject-object unity), (e) from a structure based view of the world's phenomena to a structure and function view, (f) from a reliance on knowledge derived from intellect's reason to the knowing derived from a balanced, whole intelligence—intellect, intuition, instinct, senses, feeling, insight, nurture, love, and revelation, (g) from the Earth as an infinite resource for our mere desires to Earth as the finite source for our indispensable, primary needs, (h) from the materialistic development of economic societies (nations) to the nurtured growth-fulfillment of families and communities, (i) from infinite, human development, expansion, and transmutation to the growth-fulfillment-perpetuation of our magnificent human nature, spirit, being and reality within our source, Earthlifeworld, (j) from sanity as coping and adapting to alien, denurturing, dehumanizing phenomena to sanity as growth-fulfillment of our evolved natural, human being and purpose within living nature's larger world (field) of living being and purpose, (k) from the pursuit of affluence and luxury to the pursuit of growth-fulfillment sufficiency, (l) from it takes a therapized, nuclear family to it takes a village (in harmony with living nature), (m) from individualism and egoism to human being as bio-self, family-self, community-self intergrown within and part of living nature's world (field) of beings.

In addition to the above conceptual shifts, optimism sprouts from three often hidden realities: (a) the mother of natural, human invention (as opposed to power-wealth driven invention), authentic, human necessity and needs, is on our side as we employ old and innovate new resistances against alien inventions and automatized, historical forces toward recoveries and preservations of natural, authentic humankind and nurtureculture; (b) the need to be human—the drive toward human being and humanness—is intergrown with and works with the need and drive for mere survival, the will to live. That is to say, we have more than life-force; we have within us our lifebeing-force (of human being), as in turn, the primates and mammals have theirs. Our essence is to live *and* be: to live *human being*. At times, being-force even overrides the life-force; hence, the declaration or sentiment, *Give us freedom or give us death*! Ultimately, the need is for the freedom to strive toward full human being. (c) It is heartening, moreover, that the human revelation is spreading, characteristically right in time with increasing dangers to our humanness.

Therefore, behold, believe, resist, rebel, rejoice! Once again, there are fresh movements to overcome and transcend sci-technics and commercialism! Once again, that stirring within; still again, we move homeward, homeward! Again our human being and spirit, and our human nature and reality are regathering with our natural, authentic selflifeworld and nurtureculture!

In the end we recover and love ourselves, our being. And, endings can be beginnings. A generation or an epoch grows old; another throbs with youthful spirit, purpose and instinctive wisdom, less adulterated and self-violated. This, is our way of human recovery and redemption: authentic humanity resprouting amidst the alienism of 'civilization.' This is the perennial element of the human springtime.

There must be and there has commenced a fundamental change in Western thinking, life and consciousness: a change effected by the resurging lifebeing-force of our Last Human Spring. The conduct of our personal and social lives, must be rooted in a regenerational, organismic, ecologic consciousness embracing what we as humans are evolved to be through our inherited genes and *normally inherited* nurtureculture. The latter of these is incessantly swept away through adulteration by artificial, alien elements-phenomena of sci-technic 'civilization'. Our redemption is a philosophy—or, more precisely, a selflifeworld view and consciousness—that fosters a renurturalization of our nature-intended, evolved, ecological, whole self—which embraces the family-self and community-self—and of our selflifeworld within Earthlifeworld. And it must root us in a nurturing force (overriding alien forces), as growth-fulfillment-perpetuation guidance for education, media, leisure, business and government. For, the modern West has had it backwards. We as egoself persons, as transmuted, alienized families and communities have been reacting to the self-propelled (automatized) forces of sci-technics and commercialism that direct public and private institutions and corporations. We are now entering an epoch where what is critically needed—and is ocurring—is a sustained, long-running counter-revolution of human being, spirit and consciousness. Our sprouting commandment is to subdue and subordinate the forces we have been reacting to and make them responsible to us, to our humanness.

Individualism has fallen short as human doctrine; it is a partial truth (*See* "Individualism," Part 32 of Ch. X.). It is true that one measure of ideas, actions and movements should be how they affect the individual, his mind and personal life. But, the individual's thoughts and actions are more truely and humanly measured beyond their effects upon himself and his individual (organism) welfare: by their effects upon the welfare, structure, function, growth, fulfillment and perpetuation of the family, community, the natural nurtureculture (socioculture), as well as, Earthlifeworld. That is to say, the individual is intergrown and interfunctional with all these in the organismicity of human being thereby attained. If the onus is on the individual, it is only to distribute and share the onus over into the family, community and lifeworld nurtureculture. He is to play himself, his bio-self being, within the organismicity, the ecosystems of human being and reality that includes family being, community being and its natural, Earth's lifeworld being. The individual cannot experience (nor be) whole human being in himself alone.

This requires from the West in the coming epoch: first, that individuals come together and unite with others that realize that the individual and individualism is incomplete as a human being and doctrine, respectively, and to rebuild, recover, protect and preserve family, community, nurtureculture and Earthlifeworld (the "environment").

Second, human preservation cannot be accomplished without a coming together in enclaves and groups (geographical, ideological, spiritual, and communal) to resist the artificialization and alienization that denurtures their personal lives, their children, families, communities and nurtureculture. Our ancient humanness can be defended peacefully with family and community rejections of alienization and regional offensives of human renurturalization. The Old Order Amish communities of North America provide one successful, stark model of staunch resistance to artificialization and adulteration of our natural humanness. And, many more peoples (cultures and subcultures) including some American Indians and diverse peoples scattered across the Earth serve us well as successful models for human preservation. It is up to each and all of us to determine how much resistance we are capable of, and to decide how much recovery, rebuilding and preservation of whole self, family, community and selflifeworld that should be aspired to and strived for–as well as, the actions and methods to be used. (This book will not attempt to be a handbook for renurturalization.)

The Old Order Amish communities long ago drew a line in the sand of time excluding the telephone, automobile, electric lights, secondary education, commercialism and modern clothing fashions: basic rejections that proved successful for them. All others who lay claim to natural, authentic human being, spirit and reality, all who embrace our primordial, proven, tested humanness, free of alienizing adulteration, must spend long hours and times in their lives evaluating the historical terrain–particularly the scitechnics, commercialism and their electronic media–and the happenings around them, drawing lines, setting boundaries, erecting fences that uphold and protect the humanness that scitechnics and commercialism deem inferior and replacable by career fueled, automatized systems employing experimentation, exploration, and curiosity, unrooted from authentic, human growth, fulfillment, and perpetuation; and ultimately driven by the desire and addiction for the power and wealth of scitechnic commercialism.

Is there a human revelation at the heart or center of the fundamental human revelations we can receive, basic revelations about life and being that, in turn, have smaller, human revelations springing from them? If so, such a core revelation would be the natural, authentic source of all revelatory awareness about natural, authentic, human nature, being, spirit, identity, reality and human selflifeworld (*See* "The Root Human Birthright," Part 3 of Ch. II and "The Revelation: Nurtureculture and Nurturome," Part 10 of Ch. III.). The philosophy presented herein, human authenticism, affirms and describes such a mother revelation, a regenerational awareness and force countering the human decay, fragmentation, adulteration, artificialization, alienization and oppression that the science and technics of 'civilization' and their corporations, institutions, governments, politics, media and alienizing entertainments inflict upon the natural, human being, spirit and purpose that are original and authentic to us.

Some readers have experienced parts of the human revelation; a smaller group have received most of it and are on the threshold.

Likely, a few have received it, unwilling or unable as a rule to articulate it. (Recent surveys indicate that about 10% of Americans have made a key paradigm shift.) Hopefully, this volume's articulation of it will be helpful, however imperfect it must be due to the inescapable element of subjectivity, the limitations of the author, the limitations of language and written language, and due to the background heterogeneity of the readership. For the assorted readers: (a) for those lost in the long, darkening byways of 'civilization' socioculture, (b) for those somewhere on the road back homeward via the Last Human Spring, and (c) for those on the threshold of the human revelation; all either have or don't have enough remaining, natural intelligence, consciousness, force of human being and purpose, and the personal predilection to open their conciousness to the human revelation. Regardless, parts of this core human revelation *will continue coming* to growing numbers—of those within Western 'civilization' and other sci-technic and commercialistic sociocultures–feeding the regeneration of natural, authentic, real, human spirit, being, living nature and reality. Whatever the precise outcome of this old, great, and last, human struggle, we need not doubt that diverse enclave communities, families and persons of our natural, authentic humanness will sustain themselves, and feed other quickenings of the Last Human Spring.

Have there been, then, throughout history a handful of eminent philosophical people, such as Jean Jacques Rousseau (1712–1778), the father of philosophical romanticism; Lao-Tze; the author(s) and fathers of *Chaung-tzu* (written about 2400 years ago) of philosophical Chaung-tzu Taoism; the figure called Buddha of Buddhism; and other immortalized people of thought or action that have received this core, human revelation (in a form fitting their personal destiny in a particular socioculture) and have utilized it to create their regenerational, naturalistic, thought systems or life achievements? And, have there been uncounted others within 'civilization' societies, most not noticed by history that received this core, regenerational revelation about human nature and humankind? Such a supreme human revelation, as this volume affirms, does indeed exist beneath the overlaying systems of thought, consciousness and (socio)culture of the various 'civilizations'. As stated above, a growing number are receiving parts of the human revelation.

Rebel spirits everywhere in the West can identify with the lyrics of Billy Joel's "River of Dreams:" "I go searching for something taken out of my soul, something I'd never loose, something somebody stole." Those that have received major parts of the human revelation are quite notably found amidst the deep ecologists, social ecologists, ecophilosophers, naturalists, ecofeminists, counter culturalists, transpersonal psychologists, and others in the Green Movement.

A new epoch has dawned in the West: an epoch of the Last Human Spring. It is of greater importance than the Enlightenment and the Reformation combined, greater even than the resulting emergence and spread of philosophical romanticism. We live in the most pivital epoch of history. One that will experience an unprecidented human decimation. The West has been meeting challenges to and influences upon Western human values, spirit and ethos that must and will eventually transform it and all the sociocultures under its influence. Some of the people within a'civilization' Earth-folk sociocultures sense that a small number of people within Western 'civilization' lifeway are upon the threshold of a breakthrough, a redeeming revelation that could eventually preserve the human species. Things are quickening as the light and warmth of our authentic, human spirit, nature and being counterreact to denurturalization and alienization. The last human spring is well under way, a renewed hope and self-affirmation within a'civilization' Earth-folk cultures, a reaffirmation within 'civilization' sociocultures of low-tech culture, ecologically sustainable enterprises, bioregionalism, ecosystem projects, cottage industries, local arts and crafts, a widespread movement to salvage, rebuild, or build and preserve organic, nurturing families and communities; and, of course, that ever-present, often unnoticed, motly minority that rejects material wealth and 'success' in favor of more simple, wholesome and fulfilling lifestyles considered 'old fashioned'. These developments make resignation to a false, 'inevitable' 'progress' a myth of industrialism.

Since the 1960s, an enlightened, regenerational, enlarged minority in the West (the remaining core of the 'counterculture,' post modernists, deep ecologists, naturalists, ecophilosophers, ecopsychologists, ecofeminists, transpersonal psychologists, sociobiologists, ethologists and others) have been realizing and renouncing basic, erroneous ideas about what we are as human beings, and commencing the recovery of our natural, authentic human condition through the significant retransmutation of alienized being back into human being, toward a reauthenticization–a renurturalization and renaturalization–of Western 'civilization'. (*See* "THE REVELATION:", Ch. III, and throughout.)

Human revelations, distinct from religious revelations, reveal to us profound truth about human nature, being, spirit, identity, reality and the selflifeworld of these. Human revelations can also reveal the relationship of these human sphere-elements to the overall, pervasive spirit and being-force of life itself. Religious revelations, by contrast, reveal something of the living nature of overall living Spirit or some conceptualized, creator Being. Human revelations sometimes are interpreted into a religious belief system and judged divine within the context of a religion.

These revelations provide a variety of illuminations. The paradigm changes mentioned in Part I above are quite representative, as are the revealed features of human nature, spirit and being in dormancy and suppression amidst the chill of 'civilization.' But, there is one complete, human, root revelation that floods light into every primary element of the human condition or realm. It most often comes as an ongoing series of revelations lasting about thirty minutes to an hour; but, it can occur partialy and as parts of a serialized reawakening and reconversion. It reveals the human condition, human life and being, in its natural, archetypal, authentic state or condition of nurtureculture and nuturome (*See*

Part 10 of "THE REVELATION," Ch. III, and Part 3 of Ch. II.), the way people are before various cultures of 'civilization' fragment, diminish, surrogate and adulterate them with material, spiritual and social pollution, i.e., with the alien and alien, artificial elements-phenomena that accrue in any 'civilization', and especially in the West and other industrialized societies. The human revelation rarely, if ever, comes in complete purity; for, it is unavoidably filtered and obstructed in some degree by subjectivity, ethnicentricity, and 'civilization' centricity.

## 2. One Last Respringing

A human spring, then, metaphorically and analagously, is the requickening and resurging of human nature, being, spirit, reality and their selflifeworld, all natural and authentic to us as natural living beings, a requickening amidst the oppressive, debilitating, decayous socioculture (society) of alienlized, artificialized 'civilization'. We are entering an epoch wherein the West will be renaturalized and renurturalized: its sci-technics and commercialism domesticated and subordinated to our natural, authentic humanness. It is necessary that this Last Human Spring triumph and reverse the long, progressing conditions of human decay on the personal, family, community, sociocultural and global levels. Toward avoiding human decimation or extinction, it is necessary that each 'civilization'—especially the West and the nations competing socially and economically with it—de-alienizes and re-nurturalizes back into more natural forms of human socioculture.

Overpopulation is merely the most widely recognized, alienizing element-phenomenon and problem afflicting humanity. Unfortunately, overpopulation—and the religious and political positions taken on it—distracts Western awareness away from the more basic alienization that is the growing cancer within the naturally prosperous, post-industrial societies. To recognize the original, inherent alienism, the malignancy within 'civilization' itself, is the ultimate, human necessity, now and in the coming epoch, because sci-technic 'civilization' (technocracy) has generated an accelerative alienization and artificialization of an intensity that forebodes an unprecedented, multi-agented decimation.

For Western humanity this resurgence of light and warmth is especially overdue. The two World Wars, unfortunately, did delay our spiritual and ecological regeneration. Our regeneration was overridden by survival itself; survival, moreover, amidst unhuman, alien forces of debilitation and death. Misfortunate, historical forces pushed sci-technics and commercialism into an unprecedented domination of selflifeworld and Earthlifeworld. How desperately we need this respringing of human being's natural, authentic growth-fulfillment-perpetuation of its selflifeworld! This likely will be the last (and triumphant) re-emerging and spreading of the revelation of natural, human spirit, being, nature and reality, which is authentic to our *50,000* year old subspecies. This time, hope-fully, the light and warmth will gain enough strength to hold us and stay with us to permanently check and subordinate the alien-ism of 'civilization'. This time the human revelation and the hu-man spring will triumph—albeit our triumph can only be a check on alienization, a balance of power between alienism and human being's lifebeing-force. The time is ripe for our reembrace of natu-ral, authentic human being, simply because the realizations, com-mitments and priorities that make up the human revelation are for the first time in history absolutely necessary, not just to avoid the various disasters and scourges that are looming over us, but to avoid the extinction or decimation of our human species itself.

Upon one of its recent, major respringings to Western 'civili-zation' humanity, i.e., the emergence and spread of philosophical romanticism, it was not an absolute necessity that human authenticism, our natural humanness, triumph over its opposi-tion. For then, we had time and resources to play with, to play the games of adventure-discovery, of subjugation-exploitation, of luxury-waste, of science-curiosity, of technology-challenge of egoself-greed and others. The opposition forces, entrenched against our human spirit and against our major resisting philosophy of ro-

manticism for over 200 years now, have maintained sufficient credibility to generally capture, control and predominate Western social and cultural values, willful action and futuristic endeavor. The
strength of these forces opposing philosophical romanticism and
romanticism's own flaws kept us from breaking freely and fully
into the human revelation. The situation has been much like that
of an Alaskan sun having dawned but unable to rise. Our inner
source of light and warmth, our human organismicity, was
thwarted, preventing a fuller regeneration of the spiritual-ontological vitality, natural and authentic to our human species and
subspecies.

As the redawning and resurging of human spirit, being and
nature finally subordinates alienization and artificialization to
rehumanization and renurturalization, during the coming century,
no appologies or compensations are owed to childish, adolescent,
alienized cultures of 'civilization' for the resulting curtailment of
their irresponsible play. The injurious, sociocultural games they
have played with the lives of their own people, the lives of subjugated, misled or beguiled peoples and the countless living species of the living Earth, home of all life, have to be halted both
for the sake of human life and welfare, and all life within
Earthlifeworld (ecosphere). Humankind has endured for too
long an ultimately sociocultural stage of rebellious, juvenile,
delinquent, sociocultural behavior coming to a head in the
U.S.A., Western Europe, Japan and the military industrial elements of the predisintegration Soviet Union. The
reconfigurating Soviet world has learned one politico-socio-eco-
nomic lesson, with more altogether different kinds of lessons
still to come to it and to all childish, delinquent sociocultures
now following the West's flight from human reality and being
rooted in Earthlifeworld.

This respringing of the natural, authentic, human condition–
human authenticism; this respringing of human being's natural,
authentic growth-fulfillment-perpetuation in the West comes from
without, as well as, from within its deteriorating foundations and

walls. From within, the family–which circulates social heartblood both to individual, biological organisms (mind/bodies) and to the community and/or extended family and/or clan–this organic family and, resultingly, the organic community (natural and authentic to us) are being diminished, surrogated, adulterated and disintegrated. Lacking a human revelation healing and rehabilitating our human being and its selflifeworld, the fate of organic, whole family will continue to be the same as the fate of organic, whole self and organic community: all face an increasing decay of their collective, organismic human being.

From without, the West is increasingly being contrasted unfavorably to other 'civilizations' and Earth-folk cultures less destructive to themselves, to others and to humankind's collective home, Earthlifeworld. Further, there are those still generally outside any sci-technic, urban, 'civilization' lifeway (whether Western or non-Western) that are increasingly aware of such results of sci-technic 'civilization' as World Wars, urban disintegration of the person, family and community, looming nuclear doomsday, looming bacteriologic, anarchistic, terroristic doom, sci-technic and environmental dooms; such people are displaying an increasing resistance to personal and social debilitation from alienism (traditionally we have mistaked these as manisfestations of mere *alienation*). They are in a movement of revolt and rejection of science-technologyism, and ultimately the degenerating spirit-ethos of 'civilization' itself–power-wealth. (*See* power-wealth in Ch. II, Ch. VII and throughout.) The alternative, original spirit-ethos natural and authentic to a 'civilization' Earth-folk socioculture–sufficiency-growth-perpetuation–is still respringing and pulling 'civilization' humankind back into human reauthenticization and renurturalization (*See* "THE REVELATION:", Ch. III.). The West will either cultivate this Last Human Spring, or rot with its followers in its own progressing decay, in its own sci-technic, commercial fever, addiction, and madness.

## 3. The Root, Human Birthright

Is there a root, nourishing, human birthright, sprouting from living nature's being and spirit, from which all other birthrights spring? Yes, it is the right to grow and fulfill our natural, human being (beingness), authentic to our human species: the right to become whole human beings, possessing ripening, socio-biologic-pycho-spiritual personhood (organism), family, and community. This natural, evolved human being has two sub-ecosystems (programs) evolved by natural humanity—parts intergrown and interfunctional as one ecosystem: (1) our genetic *genome* for biologic growth-fulfillment—of our mind-body organisms—and (2) our inherited (*normally*, *naturally* and *originally* inherited), nurturecultural *nurturome* for eco, personal, family and community nurturings—our nurturecultural growth-fulfillment-perpetuation of human being, spirit, and soul, within Earthlifeworld.

These two natural, intergrown, human, evolved ecosystems are intergrown and interfunctional with a third, all encompassing, ecosystem, the whole ecosphere of living nature—Earthlifeworld. The two subsystems are an natural part of living nature's organismicity. These three ecosystems interfunction as the ecosystem of natural, human being; a ecosystem-onto-web that carries the person as whole, human being (body-mind-personality-family-community) and the species through Earth-time and being. The first, the genetic design, is composed to fit the second, natural *infant and childhood nurturing, the natural social and material environment and the entire socioculture: the postbirth nurtureculture and nurturome that the species has inherited and passes on.* The exception being 'civilization' where quasi, transmuted socioculture and quasi nurturing present adulteration, fragmentation and artificializion and, are thereby, substantially alien to the genetically based social and sociocultural human purposes and destiny. In other words, *what comes after birth—mother-nurture, family-nurture, community-nurture and sociocultural growth-fulfillment-perpetuation—is normally, naturally and originally held to a correct fitting, in harmony with (in compliance with), genetic instruction; because natural selection, as well as natural will and judgement, selects nurtureculture activities and behaviors (objects-phenomena) that fulfill this proven ecosystem ontoweb (or matrix) for the species, its communities, its evolved, ecological purposes of being and its place in the ecosystems of human life and other beings–going beyond mere survival and individual reproduction.*

In other words, our most rooted birthright is the right of the human (a) mind-body organism ('individual' or bio-self), (b) family portion and (c) community portion of the human being to become what they are designed by evolved life to become, which can only happen in accordance with the two natural ecosystems–genetic, genome, and nurturecultural, nurturome–for the growth-fulfillment-perpetuation of these two portions of whole self that are intergrown within the third, naturally evolved, ecological Earthlifeworld (ecosphere). The three constitute triecological, human selflifeworld. Emphatically, natural, social beings, both animal and human-animal, grow, fulfill and perpetuate themselves with a natural socioculture (nurtureculture and nurturome); but, emerged 'civilization' substantially opposes, stunts and transmutes this growth-fulfillment and perpetuation, this central birthright, to whatever degree it has become adulterated, artificialized and alienized. Our birthright is to achieve our selflifeworld within Earthlifeworld (ecosphere).

Upon our understanding of which are those parts of human nature and being (objects, behavior and activities) that arise from genetic living nature and instruction; and, then upon understanding how mother-nurture, family-nurture, community-nurture and Earthlifeworld-nurture all function organismically together as human nature's nurturecultural nature–human nurturome–(for natural human socioculture) and intergrow organismically with biologic nature–human genome–to permit and guide our triecological, whole, human nature, being, activities and behavior to be grown, fulfilled and perpetuated; we then have the archetypal ecosystem-onto-web for human being – the human nurturome. We can then test or evaluate social and sociocultural behavior and objects-phenomena as to how well they fit or adhere to what we thereby understand is our natural, authentic, human destiny and purpose within Earthlifeworld. We are then able to begin to reform–retransmute–alienized, artificialized elements-phenomena (things, activities and behavior) of 'civilization's quasi, alienized persons, families, communities, institutions, organizations, and cultures back into a better alignment with the natural, authentic, human models of these given from *living nature's* genes and evolved nurtureculture, our nurturome.

## 4. Human Nature, Patterns, and Nutrients

We bear within us what preceded the anomaly of 'civilization', i.e., the psychological, social, and spiritual-ontological features of pure human nature and being. Our evolved human nature attempts to direct us into the organismic growth-fulfillment and perpetuation of a selflifeworld authentic to us. However, in substantial opposition to this right, fitting, good and beautiful growth-fulfillment and perpetuation are the plans and activities of 'civilization' sociocultures that incorporate alien, artificial objects and phenomena which fragment, disintegrate, diminish, surrogate, adulterate, eliminate, delay and unsequence the manifold unfoldings that make up the magnificent, human condition, the selflifeworld, that is natural, original and authentic to us.

The inherent, species nutrients of socio-biologic-psycho-spiritual, human being have been the organic family-community and the natural Earthlifeworld (commonly termed *environment, biosphere,* or *ecosphere*). The biologic 'person'–bio-self or mind-body organism–intergrows with organic family and community when these three are in their natural, unadulterated forms, i.e., within natural nurtureculture (*See* Ch. III, Part 10, "The Revelation of Nurtureculture and Nurturome.") and natural Earthlifeworld. The bio-self fulfills and perpetuates human being through living nature's ecosystem evolved for us. We, as individual mind-body organisms,

cannot have complete and normal growth, experience fully blos-
soming human being nor live as complete, natural, authentic,
human beings when fallen substantially out of these three growth
mediums. For, we each and all then become intellect-egoselves
(*See* "EGOSELF, ORIGINS, DEVELOPMENT and ARISINGS,"
Ch. IV.) having quasi, adulterated, artificialized, alienized life and
world. Living nature has created humans and other social animals
as natural, intergrown, interfunctional, whole selflifeworld; that
is, as natural bio-ego-self, social family-self and community-self
intergrown as a whole, sociobiologic triecosystem of whole self (*See*
"The Revelation of Triecologic Whole Self," Part 14 of Ch. III.),
which is in turn intergrown with Earthlifeworld (consisting of other
natural biological and sociobiological life, and with certain, en-
listed, participating, natural, inanimate, biospheric matter). All
together these constitute organismic, human species', triecologic
selflifeworld, within Earthlifeworld. A third triecosystem, body/
mind/spirit, intergrows within the first two making a tri-tri-
triecosystem of human being.

## 5. The Loss of Being

The deeper human life and being move into 'civilization'
socioculture, the more need there is for awareness of human reality
and being, as well as, for science, metaphysics and ontology. For, it
is within sci-technics and commercialism of 'civilization'
socioculture that humans experience the adulteration, both from
alien and alien-artificial objects-phenomena that diminishes human being and human life, albeit substantially unnoticed and suppressed. For, lacking any physical, mental or sociocultural impair-
ments (as social, economic or material), humans assume that they
are living a full life; that is, they assume this up to the point where
a sense of meaninglessness or unfulfillment emerges into awareness. Whereupon, they turn to false, deleterious measures in an
attempt to bring back or enliven what is missing of their
Selflifeworld, their being, and spirit. Thus, it is within 'civiliza-

tion' that human life (existence) and human being are no longer one and the same, that the schism begins, bringing progressive disfunction, conflict and pathogenesis to persons, families, communities and nations, and 'civilizations'.

## 6. Emergent, Subliminal Self-destruction

A failure of the West to experience the human revelation and redemption of consciousness would indicate the coming extinction or decimation of the human species. Failure to act upon the received human revelation would indicate the presence of a subliminal, human will to (willingness for or acquiescence to) extinction. Such a human species death wish, preceding the human revelation and conversion, has already evidently penetrated Western, 'civilization' socioculture. Among the large body of evidence is the high incidence of adolescent and adult suicide, criminal violence, and by the intensifying self-destructive personal and collective practices well known to the inner cities and suburbs of sci-technic commercial 'civilization' sociocultures, as well as, by the widespread lack of care about securing the means for future generations' survival and welfare.

## 7. Truth Vs. Human Truth

The term denoting the system of philosophy, herein, human authenticism, affirms a vital truth and distinction commonly overlooked: namely, that just as the truth is not the same as the human truth, authenticity is not the same as human authenticity–human authenticism. And, further, when elements of truth unable to participate in human welfare and fulfillment are nevertheless incorporated into human thought and activity, these alien truths dilute, adulterate, diminish and fragment human lives and welfare. We waste time and being through adulterated, alienized thought and activity not generally true and authentic to our human reality. We commonly overvalue truth itself: it is merely secondary. The

primary truth (and truths) is that which belongs within and participates in the natural organismicity of natural selflifeworld–the
human condition or realm that is natural, real and authentic to our
50,000 year old human subspecies. What primarily matters are the
elements-phenomena participating in our, own, natural, authentic
condition of human authenticism manifested in both of our natural,
human sociocultures–pre'civilization' socioculture and a'civilization',
Earth-folk nurtureculture.

## 8. An Allegory of Transmutation

Our natural, long-running human nature, being, and reality
are mixed in with and polluted by the alienism of 'civilization'.
Imagine, reader, that a Earth-folk village community experienced
a serious pollution of its water supply that resulted in progressing
mental and physical damage to its members. Suppose, also, that
the pollutant, not being evident to the senses nor intelligence,
remained unknown; and, consequently, the decline in personal,
family and community health and well being was given various,
false explanations and interpretations which were passed on through
the generations. Imagine next that this community adapted to
this human personal and social adulteration by ingenious responses
and adaptive feats including the exacerbative, the ameliorative and
the ego gratifying. Imagine these were seen to be human triumph
in the face of huge–but normal–human problems; when, in reality, they were treatments being utilized against what was, unbe-
knownst, a growing, alien cancer. Imagine, also, that this community was in a land of other self-sufficient village communities–
either settled agrarian or hunter-gatherer communities–with limited contacts for the purposes of seasonal trading and mate selections. Imagine that this community, consequently, evolved a distict
type of socioculture regarded by all other surrounding village
sociocultures, as more notable for its impressive, coping artifice
and mechanisms, than for its largely hidden physical, mental, social and spiritual, deleterious anomolies.

Imagine, reader, that consequent, sociocultural decline, hidden inside its illustrious artifice, resulted in its loss of community (and sociocultural) sufficiency, requiring (as the mother of invention is necessity) aggressive, intimidating, unfair trading practices and then raids, using invented warfare weapons, for pillage, destruction or subjugation of neighboring communities. Imagine that the original, natural purpose and motivation for aggression–defense of family-community survival and growth-fulfillment-sufficiency–was replaced, partly through the aggressor's fall into family-community insufficiency and partly through the very presence of the weaponry means, by a new, unnaturally motivated aggression, i.e., the will toward personal and collective power and wealth.

Now, reader, in the above allegory make the following substitutes: 'civilization' socioculture in the place of the village community; alien, artificial elements-phenomena (material, social and sociocultural) of 'civilization' in place of the unrecognized water pollutant. Thereby, what is revealed is the nature of the phenomenon of 'civilization' and the deceptive and incremental manner of its original penetration into the natural, human socioculture authentic to our, 50,000 year old subspecies.

Had the village community been able to realize its decline and the nature of it, it could have attained a source of water within human requirement and boundary–within the nature of human drinking water; or could have moved the community to a location with unadulterated water. The village community could have avoided becoming an anomalous village community, moving with progressive debilitation into abnormality and eventual transmutation of the human socioculture and lifeworld, natural and authentic to its subspecies and species living nature and being.

We need to realize that 'civilization' is alien transmutation is not the natural movement of human socioculture into a new sociocultural experiment of evolving living nature and human nature. Since this socioculture is one of artifice–alien artificiality in particular–(and one of increasing artificiality with some interrupting reversals effected by resurging human spirit), this is unnatural, alien human transmutation. It moves humans and their natural sociocultures increasingly outside of living nature, human nature, spririt, being, reality, nurtureculture and human selflifeworld.

Like a village acting to clean its polluted water, human, alienized cultures of 'civilization' can now, with the revelation that exposes their reality as one of adulteration, transmutation, addiction and malignancy via the alien elements-phenomena of 'civilization', begin renurturalization, renaturalization, and recovery. This is the movement out of alien, artificial elements-phenomena back into natural elements-phenomena that are authentic and rightfully belonging to/with our 50,000 year old subspecies, rightfully belonging to/with our grandchildren and a hundred generations of grandchildren. This is the movement out of 'civilization' socioculture, into a'civilization', Earth-folk socioculture of diverse cultures–the reauthenticization, renurturalization, and renaturalization–of humankind.

## 9. Three Alien Elements to Human Nature and Being

While artificiality is the most destructive alien to human being, nature, spirit, reality and socioculture, we in the modern age increasingly face an overload of natural aliens that is also destructive to our natural, species' human being. This nonartificial alienism–existing within living nature itself–can be divided into two kinds. The first is living nature's own built in, benign mechanism that consists of the aliens (foreign, mutant genes) that she produces and holds at the perimeter of the nature of each species until some are required for the adaptive change, to meet living nature's required changes of each species. (The average person has about three to five mutant genes.) This might be called living nature's experimental or adaptational alienism or foreignism. The second type of usually benign, natural foreign-aliens (within living nature) are displacement aliens. They come penetrating through the boundaries of a species' nature, coming from either the past of a species evolutionary nature (an evolutionary throwback) or coming as an evolutionary addition to effect adaptation to changing nurtureculture or natural environment (ecosphere). Strictly speaking, these are not aliens–from outside living nature–until they are affected and increased by the third type of aliens.

The third type of aliens are genuine, pure aliens from outside of living nature–artificial elements-phenomena. They come from beyond living nature's boundaries, from outside a species' living nature or its ecosystem interrelations, weakening or overwhelming it, decreasing its adaptive ability to hold its own in the organismic interplay, and the ecologic balance of aggragate species' living natures and natural elements-phenomena. Overwhelming the species' natural ability for adaptation to living nature's organismicity, they push or move the species in the direction of species' decline and extinction. What is crucial to our natural, human species and 50,000 year old subspecies is the realization that artificiality–of these artificial aliens–exacerbates the older, natural type of destructive, unstabilizing aliens. One way it does this is to increase the number and the deleterious effect of both unstabilizing, natural aliens and new artificial aliens that are able to penetrate natural, human being–conscious and subconscious. Some major examples of artificial aliens are, the microscope, telescope, television, rapid and extensive, human mobility, single parenthood, premature, deficient babies life supported by profit driven medicine and technology, and power-wealth driven industry, business, bombarding communication, advertising, entertainment and commercialism.

Alienism–alien elements-phenomena–begins transmuting our lives by meeting the needs of the acquiescent, quasi, egoself will that suppresses and is at odds with the natural, authentic will or purposes of whole human beings and their (socio)cultures. Aliens enter through the doors of our acquiescent and desirous 'will', which is in opposition to our will toward meeting the natural authentic needs of our being, spirit, nature, and nurtureculture. The superfluous wants and desires of the ego-self displace the natural, authentic needs of the whole self and selflifeworld.

It is when behavior begins to emerge out of the societal level of alienized, quasi 'civilization' that has adventured outside of nurtureculture and selflifeworld–outside the triecologic self of bio-self, family-self and community-self–that this behavior begins to meet the wants and desires (for power and wealth) of the elite people and classes controlling that level of 'civilization' beyond the person, organic family and organic community. Such behavior, therefore, opposes not only the legitimate, essential needs of humankind, i.e., those that arise from the person, his family, and community; but, additionally, bypasses the eagalitarian and democratic evaluation that originally and normally occurs when psychological and social behavior emerge at the family and community level.

## 10. The Human Break From Living Nature

Various religious and secular ideologies have portrayed the human realm as above or outside the natural realm either by divine or supernatural design, or by man's own accumulative 'cultural' achievements. However, no matter how much we become adulterated and fallen away from our natural, authentic, human nature, being, spirit, reality and selflifeworld, we can never be completely separated from these; for, to separate something from its living nature is to transmute it into a different entity, in this case, an entity created through artificialization and, therefore, an artificial entity, a non-natural, non-living entity. Since there is no such thing as artificial being, this would mean our self-destruction as living beings–our extinction through artificialization. An overriding love for the miracle of life, of plants, animals and mammalian consciousness provides some of us with the consolation that if humankind artificializes itself into an extinction, magnificent Earthlifeworld–evolving, living nature–will continue on with its dialectical dynamic of life against inanimate matter. However, everything sci-technic man touches is artificialized–adulterated–thereby damaged if not destroyed. There are now too many of us with too many agents (microscopic, megatonic, and electronic) of great destructive power for us to avoid the probability that our own extinction would include a substantial or large portion of the living species presently within Earthlifeworld. Moreover, despite the small consolation that evolving life would go on without us, isn't the loss of the human species or its decimation, the product of four million years of evolutionary creation from non-human primates, too great a sorrow and tragedy to bear contemplating!

\* \* \*

Natural organisms and natural beings are equipped through their natural evolution to struggle with, survive and perpetuate against the natural dangers and competitions they will confront in their natural environment. It no longer follows, however, that humans, as natural beings and organisms of living nature, will continue to survive until that point where, within the megaenvironment or Earthlifeworld (biosphere), a change occurs that is beyond our species' adaptation ability. We have recently lost living nature's general contract for survival given through the evolved creation of every species.

Our contract with living nature is: that natural humans innately endowed with natural, adaptive abiliites within the natural, adaptation realm of Earthlifeworld, thereby possess the human will or purposes that we are merely required to know and respect in order to adapt to living nature's purpose–the living organismicity of Earthlifeworld–in order to secure the blessings of natural, free human being. For 'civilization' humans, no longer knowing, respecting and following the laws, patterns, and paradigms of Earthlifeworld organismicity, this principle of survival that applies to all natural organisms and beings no longer applies, since modernity's quasi humans have acquired transmuted, mental faculties, social relationships and technologies: they have introduced problems that are artificial; problems that grow incrementally and are traditionally disquised as artistic and technic marvel of divinely empowered state and recently disguised as manifest destiny, and for which they have no natural endowment to deal with. 'Civilization' humans, left to deal with artificialization, can falteringly utilize solely their transmuted, artificialized, adulterated faculties and abilities.

## 11. Alienization and Artificialization as Transmutation

The more adulterated that life and 'civilization' socioculture becomes, the more need there is for the transmuted form of intelligence–intellect–that is found in the transmuted, quasi mind, quasi egoself, quasi life and quasi world of 'civilization's' transmuted, quasi human being, i.e., the intellect-egoself. In natural, authentic, human socioculture (nurtureculture and nurturome) and selflifeworld, human intelligence within the triecologic self, with its family members and its community members, is rather perfectly matched because conditions and challenges are within our species' evolved, archetypal continuity of species selflifeworld; and, the ability to live life competently is passed on from generation to generation, with few, new, unknown situations arising, very few of which are alien and none of which are artificial, alien problems. However, with the progressive accumulation of adulterating, fragmenting, alien, artificial elements-phenomena, human intelligence increasingly becomes the transmutant intellect of the quasi intellect-egoself (*See* "NATURAL, AUTHENTIC, HUMAN MIND," Ch. V.), thereby leaving the quasi family and the quasi community progressively overburdened and outmatched–they are overwhelmed. The need for extraordinary performances on the part of the transmuted, abnormal intellect, the need for superior, quasi intellects and sci-technic 'genius' increases because of the complexity of the problems which are incrementally generated within 'civilization' socioculture and explosively generated in our time. Individuals and small groups of individuals come to amass great amounts of social and material power and wealth unnatural and unauthentic to humans, and beyond the evolved capacity of ecological humans to acquire and exercise within the perimeters of natural, human, moral judgement and intelligence. Science and high-technics trample and oppress the nurture and nature of human being.

It is logical that if humankind increasingly alienizes and artificializes human selflifeworld, it must consequently, in order to meet the complex problems that alienization and artificialization produce, consequently produce an increasingly, alien, quasi, artificial 'intelligence' (which it calls intellect, to which is recently added electronic, artificial 'intelligence'), as well as, an increasingly allien mind and consciousness for its quasi intellect-egoself, quasi life and quasi world, to match this progressive alienism of the natural Earthlifeworld (ecosphere) and human selflifeworld. This can continue only to that point where human selflifeworld falls apart, its center failing to hold, i.e., to the point of extinction or decimation. By reversing this progression into artificial, alien realm— through subordinating and domesticating alienism to human authenticism's human nurturome—we achieve our partial human deliverance.

## 12. Secular and Religious Responses to Alienization

Modern humankind, i.e., our 50,000 year old human subspecies, has been pulled into 'civilization' socioculture gradually during a 5,000–10,000 year period, though living nature and human nature have repeatedly downsized or destroyed 'civilizations'. The organized religions of 'civilization' sociocultures, as opposed to pre'civilization' and a'civilization' spirituality, sprang up to explain and address the problems that 'civilization' presents for human spirituality, which is considered by them to be divine. They cultivate and preserve the sprituality that organized religions consider as part of human nature, spirit and being that are under seige by the elements/phenemena of 'civilization' (the worldliness and sins of 'civilization'). There has not, however, emerged an accepted, long-running, non-religious explanation, parallel to the religious, addressing the problems or adulterations which 'civilization' presents to what is felt by many to be a natural, pure and unfallen state of human spirit, nature, being and soul that existed before and outside of 'civilization'.

Religions curiously and conveniently neglect to give a significant portrayal of the human lifeworld that preceded 'civilization' lifeworld, and neglect, likewise, to give a portrayal of this natural, human lifeworld's human nature, spirit, being and soul. They have a vested interest in portraying a mysterious creator or creators and divine plans and purposes; for, they have come to know more about their imaginings and doctrines of these than of free, natural, human nature, spirit, being and living nature itself: they have a vested interest in the growth and power of their religious institutions and organizations. Religions assert that these religious organizations, their material monuments, and their associated cities and states of subjugation, oppression and exploitation better represent a creator or creators than the actual creation once all magnificently at hand–living nature and human nature, spirit, being and nurturecultural commmunity. Starting from this false foundation they are destined to carry their followers farther away from a union and/or integration with Earthlifeworld and creation's human nature and being.

To be sure, various secular–philosophical or scientific–activities and explanations have emerged to portray, describe or define the living nature of pre'civilization' and a'civilization' human nature, being, spirit and their selflifeworld (structure-function-dwelling); and to explain and portray how these are damaged by the elements-phenomena of adulteration that grow within 'civilization', sociocultural lifeway. Some of these have sprang from authentic, human revelation. They have all been overridden by the wealth and power amassed and wielded by religions and political ideologies through their false claim to be the legitimate representative of the creator or god(s). The latest challenger is proving to be the strongest, namely, twentieth century secular humanism which alternately competes with and cooperates with organized religions. For, while their portrayals of creation and humanity are different, their spirit-ethos is the same–power-wealth-domination.

## 13. Values and Beliefs Attain to Human Transmutation

Since values and beliefs are natural, essential, inseparable ele-ments of human reality, being, mind and spirit authentic to a human species; alien, artificial values and beliefs held by natural beings—by human beings–constitute the transmutation of hu-man reality into human unreality (illusion), of human being into alien being (alienism), of human mind into non-mind (mindless-ness), of human spirit back into prehuman, animal spirit (*de*spiritedness). All this is traditionally termed, dehumanization. (*See* HUMAN AUTHENTICISM VS ALIENISM, Ch. VIII.)

## 14. Through Sociocultural Into Species Unreality

The greater the influx of alien socioculture ('civilization') into the natural socioculture authentic to our human species and subspecies, the greater the megaillusion becomes, moving *toward* pure unreality (psychosis) which a human species cannot support or survive with. 'Civilization' is our sociocultural journey into madness. Despite breakups and rearrangements, its push is constant and accumulative toward greater unreality: a journey into human extinction or decimation. The human revelation–and respringing human authenticism–precedes the halt and reversal of this process through the final regeneration of our real, magnificent, human life-being force authentic to us. With its growing warmth and light, some of us in the West–while confining and subordinating sci-technic intellect and egoself–recover most of our authentic and real intelligence and human spirit, our soul and being. That is to say, we commence the countermovement to this cancerous experiment, modern 'civilization', this alienism: we commence the recovery movement into human reauthenticization, into renurturalization and renaturalization.

# B. Philosophical, Human Rediscovery–Human Authenticism

## 15. Human Authenticism, the Philosophy in Brief

Human authenticism, at last, is the revelatory philosophy that reveals the natural humanness authentic to us as a 50,000 year old subspecies: our human nature, being, spirit, identity, reality, selflifeworld, nurtureculture and nurturome. The natural human condition, authentic to our 50,000 year old subspecies, which Rousseau could not successfully complete a true vision and conception of; and which romanticism attempts to recapture elements of; now, human authenticism, as a complete system of philosophy reveals quite thoroughly. It reveals the condition or state of human authenticism, which is the growth-fulfillment and perpetuation of natural, authentic, human nature, being, spirit, identity, reality, and their selflifeworld (their structure-function-dwelling).

A journey back through 'civilization's nurtural and cultural mutations is a rediscovery of our original, untransmuted humanness and our nurtureculture and nurturome that fulfilled it–prior to the influx of adulterating, alien, artificializing, quasi socioculture. A rediscovery of this human being and spirit that has carried us harmoniously and splendidly for some 50,000 years, is much more exciting and rewarding than our present, harmful acceleration into an increasingly inundating, churning, fragmenting and alien future. Surely, this is the greatest and most magnificent challenge we can shoulder, to recover and secure much of what is lost and missing of our natural, authentic, human nature, being and spirit and the nurtureculture of these, for ourselves and our posterity; to protect this humanness against the disorienting, disintegrative elements of egoism, intellectualism, novelism, creativism, accelerative change, adventure-discoverism, imagination-inventionism, challenge-technologism, curiosity-sciencism, thingism (objectism), waste-luxuryism, entertainmentism, celebrityism, personal empirism, and other human adulterations; and against the decayous power-wealth spirit-ethos generating these engulfing, decaying ethicisms. The journey in this epoch back into natural, authentic, humanity and nurtureculture is our last journey, the one of deliverance for 'civilization' humankind.

The purpose of authenticism, the philosophy, is to optimally rediscover, recover and unify, for our subspecies, our persons, our families, communities and sociocultures, the original, authentic, magnificent, undiluted, nature, being, spirit, identity and reality of modern humankind, and the selflifeworld of these. The purpose is recovery: to recover much of what we were before the transmutation of whole self and lifeworld via artificialization and alienization; to reconstitute people upon their authentic, dual nature–their primordial (a) biologic and (b) triecologic, social and spiritual-ontological–nurturomic natures; this dual nature and being, sprouting from primordial Earth-being of Earthlifeworld. And further, human authenticism opposes, resists and renounces the modern movement to abandon our natural, authentic humanness via an incrementally increasing nurtural, social and cultural (sociocultural) transmutation via alien, unnatural, artificial elements-phenomena of culture, in favor of a new quasi human–a supposed, sci-technic 'superman'. The new, quasi-human species—phoney, unviable, denatured being correctly termed *Homo alienus*–taking form and to replace our natural, authentic human species is partly of the evolved nature of living nature and partly of science-technology; of living nature biologically, of sci-technic mode 'culturally'–half natural, half unnatural (alien)–synthetic, blended man.

As explained in "DIALICETICS" Ch. VI, everything that is not of living nature and human nature is opposed to them in universal theses/antitheses. Earth's evolved, living nature absorbs into her a fraction of her self-produced opposition–only that which benefits her–rejecting and defeating the bulk. In 'civilization', however, natural man is opposed by (and blended with) unnatural, 'socioculture', unauthentic and alien to his nature within Earth's living nature. And, unlike both dynamics of natural man–living nature's dynamic (evolution) and the human dynamic (*See* Ch.

VI, "DIALECTICS."), of growth-fullfillment-perpetuation of natural, authentic, human nature and being–this opposition or antithesis is not of natural ecosystem, not of living nature herself as a mechanism of perpetuation (as are natural selection, mutation and reproduction); and, therefore, is not of a natural human dynamic. For, mutation and natural selection together are a natural *part* of human nature and being, a perpetuation feature (for selection), i.e., a part of the authentic dynamic of natural, humanity. However, alien (unnatural) transmutation is not a part of evolutionary human nature and being (*See* quasi nurtural and cultural transmutation in Ch. II, Ch. VII and throughout.).

Unfortunately, there is no 'ideal' place, period or state of the human condition–a particular, natural socioculture, a society or culture–in our past or present that singly exemplifies for Western 'civilization' man, the condition or structure-function-dwelling of human authenticism that urban industrial people should now aspire to. Fortunately, as mentioned earlier, there are many, instructive models, past and contemporary for prospective families and communities to eclectically draw from. The purpose of human authenticism, the condition and the philosophy, is to resist alienism and its dynamic. In fact, their purpose is twofold: the above resistance; and the recovery and unity of elements-phenomena authentic to us, i.e., the human movement into reauthenticization, renurturalization (or renaturalization) and the resultant healing of decay or atrophy accrued within 'civilization' lifeway (alienism). Hence, it is not the purpose of this book's philosophy system, human authenticism, nor the book itself, to propose the return of humankind or communities to some 'ideal' period such as the Old Stone Age, the Neolithic Age; nor to a lifeway such as that of rural Iron Age, or Bronze Age; nor some ruralism, two, or one thousand years ago, nor a recent, particular lifeway. The purpose is to identify natural, authentic, nurtureculture-elements belonging to/with the nurturecultural nature of our subspecies–to optimally recover those lost and maintain those retained; and to promote the identification of unnatural, unauthentic, *un*nurtureculture elements alien to us as a species–remove and diminish these as rapidly and substantially as is practically feasible within stability: moving persons and social units daily, monthly, yearly and generationally, out of so-called 'civilization' lifeway and sci-technic 'civilization' lifeway, into natural, authentic human society of triecologic self and its triecologic selflifeworld. This movement and destination cannot be called 'civilization' which decreases our natural, authentic, *nurtureculture*; but, could be called human renurturalization or human reauthenticization, and human renaturalization, as well.

## 16. The Eclipse of Being and Meaning by Personal Identity

Until recently, has history ever witnessed human society so preoccupied with the question: 'what is (or what is the meaning of) human life, or what is human being?' Only in the twentieth century have there been a large and growing number of people suspended enough of their daily lives, outside of authentic, human being, to have this question frequently arise. For, until recently, there have not been enough unhuman forces diminishing the experience of authentic, human being. More precisely, until our age there have not been enough alien forces producing enough alien and unhuman phenomena alienizing enough urbanites to draw them out of the phenomenon of being to the point where they sense some loss of their human being. It is not until something is missing simultaneously with the presence of the means to formulate the contemplations of *what was it and where is it?* That the questions become a personal and social engagement; and the search, a personal and social undertaking. Just as the search for personal identity doesn't begin until it has been substantially lost–hence 'identity crisis'; so it is, also, with collective identity and personal and collective human being. Sci-technic man has a growing crisis of being and nature underlying the widespread, widely addressed, personal 'identity crisis'.

One central problem is that by focusing on finding our personal fit in a society, complicated by an explosion of artificial work and leisure activity, on being successful, and by relishing in the individual's act of living which is conspiciously a modern orientation; we, in time, lose much of our functioning knowledge, awareness and experience of the essences of ourselves as human beings, of *what* we *are*. This spiritual-ontological loss of identity as human beings is the price we pay for increasing our identity as egoselves, (career-consumer-selves); it's the price paid for our experience and appreciation of the individual, personal act of modern, urban living. The egoself can experience only a portion of our whole self's being, as well as, only a portion of lifeworld being. Herein, is a central flaw of existentialism, as well as, of alienism. Both propose that human being can be experienced and understood by the separated egoself. But, this egoself, by falling away from its natural intergrowth and interfunction with family-self and community-self, and natural lifeworld can experience whole self and being only partially. More important, the whole self is synergistic; thereby, making limited, egoself being, the same as mutant, alien being or unhuman being. Human being is present only in whole, human, triecologic self, intergrown within natural Earthlifeworld (*See* Part 13 and Part 14 of "The REVELATION," Ch. III).

## 17. Doctrine of Infinite, 'Human' Possibility vs. Fulfillment of Human Nature

Another main problem with sci-technic socioculture is that its alien, artificial creations call upon us to come forth with an alien, artificial quasi wisdom or quasi intelligence to deal with them, and the forces and phenomena they create. We have not and are unable to do so. Not just because we have not truly understood what we have done and are doing. But, also, because we have been by nature incapable of attaining some supposed, artificial wisdom that doesn't even exist. It cannot exist because the alien, artificial, quasi knowledge we use for these creations, are outside natural, authentic mind (*See* "NATURAL, AUTHENTIC, HUMAN MIND," Ch. V), understanding, being and reality and do not allow for the evolution of a wisdom or intelligence for them. Instead, we make do with crutches, partial remedies and false 'artificial intelligence' with its underlying, inadequate propositions and methods.

Put precisely, all natural beings, as authenticism reveals us to be, have finite abilities inherent from evolved genes and evolved natural nurtureculture. Humans have obtained through 'culture' some unnatural abilities–artifice–which the reigning philosophy of sci-technic socioculture–existentialism-futurism, which is more precisely *alienism*—holds, under the doctrine of infinite human possibilities, to be unlimited abilities: this alien philosophy holds that given limited, natural, human ability and (1) enough time and intention with (2) enough science and ingenius sci-technics and commercialism–artifice–personal and social human accomplishment, indeed self-creativity, self-recreation and human recreation is open-ended. The two central flaws are in existentialism: (1) its assertion that we lack a human nature. Existentialism-futurism must deny the normally obvious: that we are still (partially) natural, sociobiological beings, and still possessing as natural beings, a living species' social, biological, and spiritual-ontological nature. And it must deny (2) that modernity and progress are actually the adulteration and disintegration of our natural, human condition (selflifeworld) they have denied. These denials are necessary because the increase of unnatural, artificial ('cultural') development and ability, and resulting acquired, unnatural, artificial, behavioral patterns, *entails* an increasing violation (adulteration, diminishment and/or cessation) of natural, human abilities and resultant, natural, human behavior (and social paradigms), which in an open-ended progression entails eventual complete or pure artificial abilities, behavior and being thereof. However, artificial being is nonexistent and a contradiction; therefore, the doctrine of unlimited human possibilities is false: it ends; and, it ends at nonexistent being or human extinction.

The rosy, heroic, adventurist doctrine of the brave, challenging, exploring, discovering, creative philosophy, existentialism-futurism, is a rosy, brave, exciting movement into blended, synthetic being–diminishingly human, increasingly artificial. Such being is both alien to natural being and decayous to natural being. We must, therefore, continuously reembrace natural (socio)culture–nurtureculture–and forever resist artificial, quasi 'culture' and quasi nurture; and *reject* the illusion of increasing "open-ended" artificial 'culture'; and its illusionary doctrine of "open-ended" human transformation into a illusionary "higher" form of human being.

What wisdom, then, must we urban, sci-technic people, now seek and find? What is truth for a people, a time or an epoch often is folly for another. Most seek, as they must, the truth for their selves, for their peer group, for their time and place, for their quasi families and communities. We must too, but must surpass this. In our time and all future ages, we, at last, must seek to recover non'civilization' Earth-folk truth and wisdom constitued as three intergrowth ecologies: (a) triecologic whole self, (b) selflifeworld, intergrown with (c) Earthlifeworld: we become the seekers and rediscoverers of the true nature and soul of our species including its limits and boundaries. Our rediscoveries include much magnificent human being and reality lost to the adulteration-pollution from alienized, artificialized socioculture. For, our whole species is threatened by extinction or decimation; as well as, scourges and holocausts of old and new orders. The philosophy of human authenticism is the only revelation that can save us from these megatragedies. It is the only system of understanding that is a unified and whole view and revelation of human nature, being, and reality. The other truths of modern, sci-technic knowledge will help in the daily, yearly and regional maintenance of human life. But, when they clash with the truth that provides for human life and welfare indefinitely, these one day, one year, one generation at a time truths, must yield to the underlying permanence and centrality of our authentic, human condition, primordially-proven human authenticism.

We, of 'civilization' sci-technic humanity, now have revealed for us a system of thought, a philosophy of humankind and society that is non-reactive, that is endogenous, that is produced by recovered awareness of what and who and why we are, and by functioning there in that true nature and being. We can now release our natural, authentic spirit and being of our 50,000 year old subspecies from its subjugation by alien power-wealth spirit-ethos (*See* power-wealth in Ch. II, Ch VI and throughout.). Our being needs its structure-function-dwelling, its medium, its authentic, human selflifeworld. It will enter this through the reunified, natural, authentic, whole self. Once there, it regenerates the dynamic of our human authenticism, overwhelming, overriding and subordinating the dynamic of alienism.

## 18. The Transmutation of Human Will and Purpose

There are three distinct types of will or purpose which are commonly confused as being one and the same. They are (1) the biologic-ego-self will, (2) the transmuted, intellect-egoself will and (3) human being will or purpose(s). The first is the will of the biologic ego-self or mind-body self that wills toward satisfying biologic needs. The second wills toward mere wants and desires related to power-wealth and its artificial and alien pleasures. These first two wills are often at odds with the third kind of will that can be more correctly or aptly termed purpose or way of human being. The first and third, when in original, natural form–when unified as, authentic, human will–were as one. However, within 'civilization' socioculture, the will of the bio-egoself or mind-body organism becomes, at times, transmuted into the will of the intellect-egoself (*See* "EGOSELF, ORIGINS, DEVELOPMENTS AND ARISINGS," Ch. IV.); it becomes substantially oppositional to the will, the purposes, of human being. For, the intellect-egoself, regardless of its degree or caliber of intellect, has been separated from some portions of the whole self and from some portions of its original, fitting, authentic, natural lifeworld, which has been transmuted, fragmented and adulterated into separated quasi life and quasi world.

As this intellect-egoself procedes increasingly deeper into alienized and artificialized, quasi intellect-egoself and quasi

socioculture; human being, thereby undergoes continuing diminishment along with human spirit and reality; and, simultaneously,
the needs of this diminishing being are diminishing. Consequently,
the intellect-egoself finds itself increasing its wants and desires in
an attempt to fill the growing vacuum or emptiness left by diminishing being. The addiction to the artificials of the city has taken
hold. These wants representing the will of the intellect-egoself are
increasingly out of line with the purposes and needs of human
being. We find ourselves wanting to be entertained in leisure and
regimented in work, willing subjects as well for the values these
promulgate. There is less real life and being to experience, less to
do, less to relish and celebrate as the needs of human being increasingly become unknowen, neglected, displaced and lost in the
explosion of mind, being and socioculture effected by the accelerating penetrations from artificial, alien objects and phenomenon
into natural human selflifeworld. This alienized egoself still has
some legitimate needs, i.e., those rooted in that portion of human
being dwelling within the biologic, mind-body organism. But,
the wants of this intellect-egoself are not genuine, authentic needs
of human being because the intellect-egoself has been transmuted
along with its life and world, and is only partially real and natural,
having been partially surrogated, fragmented, disintegrated, adulterated and diminished by the alienizing, artificializing socioculture
of 'civilization'. The failure to distinguish human needs from human wants and desires becomes a mark and a basic flaw of 'civilization' socioculture.

## 19. The Last and Lasting, Human Struggle

Insofar as modern humankind now approaches extinction or
decimation through artificial 'nurture' and cultural transmutation
and faces ghastly scourges and disasters of old and new kinds; the
philosophy herein bursts upon Western humanity as the long gathering, bellowing protest and affirmation of our natural, human
spirit, being and nature; and also, as the first complete and final

conceptual system of species self-defense and preservation of both its life and its nature, being, identity, reality, and spirit.

The last struggle of our human authenticism, of our natural, authentic, organismic human being and spirit, will be our finest hour, win or lose, for all those engulfed by Western 'civilization' socioculture and industrial societies in general. Moreover, the reader can take heart that authenticism, like any true and good, natural system, assures victory through its very natural, authentic system. The breakthrough into authenticism, at long last, presents the clear realization and lucid explanation of the process and dialectic of our extinction which is underway, and which is to be successfully challenged and reversed by philosophical romanticism's successor, authenticism. In doing so, it puts forth, at long last, an acceptably thorough definition and description of modern humankind, the subspecies, and its natural, authentic society, community, family, and self. None of the above can be secured from death by natural disease, disaster and from quasi, alien nurtural and sociocultural mutation (*See* human transmutation in Ch. II and throughout.), until the wholesome, archetypal, human models from our nurturome are defined. The survival of each and all now depends on a working definition or description of *what* we are and should be as natural living beings—what our nurturome is. This definition must *work* for *us*, not for progressing *illusions* of what we are and can become. The definitional debates about man's true, fitting constitutional nature have been long-running and numerous. This quandary ends with the human nurturome. Something this profound has to be traumatic and magnificent, agonizing and joyful; like birth or rebirth. And like birth and rebirth, it requires some natural, normal self-sacrifice: of bio-self (personal, organism-self), toward growth-fulfillment as a whole human being. Moreover, it requires a general *intellect-egoself*-sacrifice. It is from the alter of sacrifice—sacrifice of our sci-technic 'civilization's collective, intellect-egoself that our redemption and preservation as a human species springs. With the rebirth of our whole, natural, human self, our human selflifeworld recovers and regenerates its growth-fulfillment-perpetuation, its lifebeing-force. We are allowed to resume our part in our human family's play (some one to four million years playing). It is our self-perpetuating play; for, we again are unified, intergrown and participating with the four billion year old play of Earthlifeworld. It is the only play in the universe we can join and belong with; for this Earthlifeworld created us.

The spiritual breakthrough has been perennially occurring, as it has been, for example, sporadically occurring within the various religions of humankind. But, the intellect–as opposed to under-standing or intelligence–has held out, incessantly pulling the frag-mented mind, self, lifeworld and being, out of unified structure-function-dwelling into fragmented, constituent, sputtering stucture/function/dwelling. Now, at last, intellectualism pushed to its zenith and dead end, is breaking back through to the natu-ral, whole intelligence of human authenticism. The Western intel-lect (and all industrialized intellects) now begins its process of redemption through the revelation, that it working alone–as the mind's oppressor–has been all along a central obstacle to complete understanding and intelligence and to complete human being and reality. What it has come to, after all, is revelation or per-ish; accept the ultimate light of the intellect revealing its true, limited nature and function, or dwell in the growing shadows of the intellect's increasing decay. The age is now or never. It is intellect-egoself abdication and rebirth of whole self and natu-ral intelligence; or it is extinction or decimation. Frightening, awesome and challenging; the crucial, urgent truth of every hu-man juncture is such.

Changing metaphors, the intellect's creations are many; the creation of the intellect's womb is one. From amidst the intellect's manifold creations–its artifices of objects, phenomena, ideas, sys-tems and philosophies–one thing special has been growing, like cells, tissue and organs in a womb, 'a womb of the intellect.' The intellect is heavy with child. It is about to reproduce not itself or an offspring; but, it will reproduce what it was mutated from. It is about to reproduce its parent or archetype. It is about to *un*mutate–

to produce natural, authentic understanding–intelligence. At our critical stage of sci-technic, human miscarriage, it is either rebirth; or it is final miscarriage, or still birth. The intellect of 'civilization' humanity has been impregnated and carried to term.

## 20. The Authentic Bears Fruit–Human Reality

The value of and movement into authenticism serves us a purpose similar to the values of goodness, truth and beauty; the greater we are of all these, the better we are: the more we have of security, wholesomeness and fulfillment. All of the three have traditionally been universally sought and valued; but, it is only of late that the *human authentic*–authentic humanness–can be increasingly conceived, sought and valued. Only when something vital becomes lost and lacking to deficiency, is it conceived as an essential at all. Do we not apprehend and reach for courage just when it is being opposed by fear; and for water when dryness is sensed? Such it is with the binary conditions and concepts, e.g., truth/falsehood, beauty/ugliness, love/hate and goodness/evil.

Human authenticism, or authentic humanness, like reality, has to decline sufficiently–to the point of deficiency–before it is conceptualized in awareness. Within the original state of pure human reality, the concepts of real and unreal cannot exist. It is only when the unreal appears within reality, that, with contrast and conflict, both are conceived. So it is, also, with authentic humanness–human authenticism. That is to say, human alienization had to appear and grow sufficiently before it and its opposite, authentic humanness, could reemerge as awarenesses and conceptions. The concept of, 'the authentic', unlike 'the real' is comparatively new. People have long resisted (in minorities) various forms of unreality, e.g., insanity, paranoia, deleterious obsessions, depression, drunkenness, hallucinations, or any orientation in conflict with, or in excess of, an original, authentic one. It is only of late that we begin the battle against the unauthentic; at a time when the battle against unreality has deepened into trench warfare. The intensified unreality of our time has now made a quantum leap into unauthenticism into alienism. And, paradoxically, this increased appearance of the unauthentic and, consequently, the authentic and their conception by the intellect, makes possible an effective blow to the ancient foes, falsehood and unreality. For, authenticism is the missing instrument and the missing conception on the entire human battlefield of 'civilization.' Authenticism, understood, embraced and utilized, secures not just the authentic; but, reality. Just as love wielded destroys hate; and, truth wielded destroys falsehood. What a surprise and a discovery! What a revelation of human deliverance!

But, there's more; this restructuring of intellectual, sci-technic understanding and consciousness–back out of intellect and power-wealth ethos, respectively, into intelligence and natural, human will and purpose–redefines as it clarifies truth, goodness, beauty, et al. The breakthrough metamorphosizes–more precisely, unmutates–'civilization' humanity, recovering its unified selflifeworld. For, if authenticism as it is outlined here is true, these strands of truth reconstitute the helix of the 'material'–the makeup "knowledge" itself–of truth as we 'civilization' people have never before been able to see it. It reconstitutes knowledge and truth back into the natural phenomena (the context) of naturally evolved, ecological, human being unfolding–growing, fulfilling and perpetuating: natural, human being ecologically participating in all of Earth's life and being as the unified organismicity of Earthlifeworld (biosphere life). And, it cannot be truth alone that is thereby reconstellated. Any philosopher, layman or professional, will tell you that truth, beauty and goodness are interwoven.

This triumphant, human revelation for 'civilization' humanity, human authenticism, delivers reality, truth, beauty and goodness, at long last purified and unified, to the intellect retransmuting it back into intelligence. The long awaited birth child of the intellect is not its offspring but a mutant of intellect: recreated, human intellience and consciousness intergrown with human spirit as human faith–human purpose. Changing metaphors, the adolescent bachelor has abandoned his lover, the power-wealth spirit-ethos, for a union with human spirit. Their child is their human redemption: growing, fulfilling, perpetuating human being.

So, authenticism, in its new restructurings, satisfies the modern liking for novelty. Play, then, with the new toy for awhile, until the novelty value is gone. For, when playtime is over, a human adventure begins—the greatest, most magnificent adventure in history, in 'civilization'—the rediscovery of human authenticism with its magnificent, nature-given and secured human being and spirit! What it was, how we lost it; what it still is for some, and how we win some of it back!

## 21. Toward Western Unity With Humankind

If, for roughly three hundred years, the West has taken more the intellectual route, while the East stayed more to the spiritual; the deliverance of the Western intellect through human authenticism provides a new basis for unity, peace and welfare within humankind. For, authenticism provides for the unity of the mind, body, and spirit by restoring the mind to wholeness, having intelligence as guidance, not intellect; and recapturing natural, authentic, human spirit and purpose, routing the alien power-wealth spirit-ethos. It provides the basis for the harmonious unity of understanding and spirituality, unlike the scientific/technological intellectualism of the West, the separated Soviet Union nations, and Japan, which has failed at human unity even without warfare. This missing unity has been one of the sources of the threat to the welfare and viability of industrial humankind. With authenticism, we can now begin to move into that unity.

"There is no army as strong as an idea whose time has come;" and, there is an idea ripened with need that simply by the strength and virtue of human need can overpower the integrated, systematized forces and ideas that have built the twin fortress of power we know as sci-technic bureaucracy and conglomerant corporation. There *seemingly* are no forces equal to that of sci-technicism;

and no force of idea equal to its force of idea; because it subverts, converts and coerces every other new or outside idea and force to serve its ends, and subjugates every new force to work within and for the fortressized system of sci-technic power-wealth.

This apparent omnipotence of sci-technic bureaucracy is illusion, however. There are two forces that are greater; self destruction is greater than any created system: creation and destruction being a cycle stronger than either part of the cycle. The second is the force of human authenticism—the growth-fulfillment and perpetuation of natural, authentic, human nature, being, spirit, identity and reality. The global, bureaucratized, sci-technic society cannot withstand the regeneration of human authenticism without yeilding to comprimise. For, there is no disease that can endure the re-emergence of the conditions of health; no serious transgression from living nature that can endure the self-restoration of living nature, the preservation of her vitality. There is no serious transgression from living nature's human nature and human being that can endure the regeneration of living nature's human nature and being; and, no unreality that can endure the re-emergence of reality; and, no darkness that can endure the spread of a reforming light source.

There is a growing, though struggling, sense in the most sensitive hearts, that humankind is upon a threshold; that some authentic, human glory is about to spring; about to, because it must. Just as spring must break winter, and morning must flood out the night. What these hearts sense is the natural, authentic human spirit, which now hovers on the horizon of 'civilization' humankind; ready to partially reclaim and redeem us. The spirit-purpose of our human species being is about to break again like returning spring upon the cold, dehumanized, alien unreality of Western 'civilization' humanity. It has been a long winter; and we, like Eskimos teased by a sun that clears the horizon only to set again, await a *rising* sun, a true dawning of an epoch of the natural, authentic, human selflifeworld and human society rightfully ours and all Earthlings by the evolved nature of human being, the condition of human authenticism.

It will continue coming. As night follows day, spring follows winter, goodness follows evil, and reality follows unreality. But not without the struggle of our life, the greatest struggle of Western humankind. The seeds of alienism, of unnatural, unauthentic elements/phenomema, multiplied and many-kinded, have been sewn and grown tall during the epochs of alienism. Nonetheless, alienism will commence its decline as it overlaps into the dawning future epochs of authenticism which will regenerate within 'civilization' the episodes of the last human spring, its human reauthenticization, renurturalization, and renaturalization.

This alien winter and its present, sci-technic, blizzard, must now begin to break. We are not Eskimos; it took millennia to make the Eskimo. We have not the time nor purpose. Humankind is humankind; not this churning, chaotic masquerade festival engulfing us. Let us receive our rising sun, our light, and begin the movement to recover and unify our natural, authentic selflifeworld, our complete nature, being, spirit and reality. It is this, or decimation, or extinction. Let us cultivate our respringing human authenticism in this last human spring. May it someday reach again its summertime.

## 22. Digging Out, Cleansing and Preserving Humankind

At the foundation, or in the center, of the modern, 50,000 year old, human subspecies, there remains human primaries: primary, nurturecultural system, primary life-world, primary whole self of primary human nature, spirit, identity, being, and reality. These belong with and are authentic and fitting to our primordial, archetypal subspecies, Homo sapiens sapiens. Upon them, oppressive, alienized, 'civilization' peoples build varieties of secondary, quasi systems, life-worlds, selves and realities, that are adjuncts and modifications, but which are perceived to be *the* structures of mankind. But these are, in reality, modified versions of the originals they rest upon; they are carried by, supported by the originals. Being dependents of the parental primordial, they should defer to and be quided by their proven, parental models. We shall, herein, term this original, archetypal, modern humanity, *natural, authentic humankind.*

Those few who have been willing to subordinate their vested interests—career and ethnocentric—and receive selflifeworld shattering enlightenment have begun to realize, through findings in anthropology, archeology, biology, sociobiology, ethology, and ecology, that primordial humans were not wilderness savages that we have been condescendingly told they were by an educated, propagandized class, but were competent, loving humans of family and community. That is, primordial humanness and human being are rightfully and fittingly, in fact, our own, although they are oppressed and suppressed by the dehumanizing, adulterating elements-phenomena of 'civilization' that are alien and artificial to us. Human authenticism offers a thorough, unified description or portrait of this natural, authentic humanness that is our birthright and rightful heritage, but that has been substantially taken from us through our increasing, adulterating imprisonment within sci-technic socioculture. This real, authentic humanness of Homo sapiens sapiens, lives within us inside the phony, artificial, synthetic, accumulated overlayings of false socioculture, unauthentic and alien to us. The distance we as persons, communities and societies have moved into phony inauthentic, transmutant quasi nurture and quasi socioculture is the measure or degree of the unreality of our self, life and world (selflifeworld). We are real, natural, authentic human beings by this measure: how much we have retained of our natural, authentic, unified selflifeworld, with its authentic nurtureculture, all interfunctioning within Earthlifeworld, ecosphere.

Historical people—ancient, medieval, and especially modern—are the adjunctive offspring of natural, authentic humankind. Human memory, outdistanced by time, misplaced in alien place and barraged by historical and contemporary perception, has substantially lost account of its natural, authentic humanness, excepting largely for the instinctive self. Adopted by surrogate, adjunctive systems, quasi lifeworlds, selves, and realities, at times obedient, at times rebellious against alienization; 'civilization' humanity through the consciousness of naturalistic science and philosophy has been uncovering its lost parentage, though still resists the full recognition of its authentic identity, nature, spirit, and being. This resistance, of course, arises partly from the entrenched, stereotypical view that pre-'civilization' man was dumb, inhuman, apish and savage. The choice confronting historic humankind for the very first time parallels the one faced by increasing numbers of modern individuals: to let things be as they have turned out; or to make contact with, associate with, or even rejoin his natural, authentic parentage. The analogy ends here. For sci-technic humanity has but one choice tenable with perpetuation and welfare. Now living within alien, adulterated physical, social, psychological, and spiritual systems, lifeworlds, egoselves, and quasi realities that are increasingly destructive of the nature, being, reality and consciousness authentic to us; and that guarantee our decimation through megascourges, disease, toxication, and nurtural, cultural transmutation; 'civilization' humanity, long astray and now in growing darkness, on the edge of a final abyss, must move toward a long lost home, to our evolved nurtureculture and selflifeworld, i.e., to authentic human nature, being, spirit and reality.

It seems difficult or impossible. How much easier, seemingly, it is just to hope for and yield to a species extinction of ourselves, coupled with some conceived, promoted afterlife. In a few generations it will be over and done, if we, this human species, temporarily survive the nurtural and sociocultural transmutation into a new quasi species or subspecies. 50,000 years seems like a long time for our natural subspecies to have lived; though our predecessor, Neanderthal man, made it for 300,000 years or so. However difficult the defense and preservation of modern humankind is at this late stage; we should remember that self defense is a tradition of all life; and on the species level we are the first that is capable of recognizing that the process of our extinction is underway, of understanding it and fighting for our survival and perpetuation. Then, too, a slow surrender to extinction or decimation means a lingering life with only alien pleasures and comforts, without dignity, outside any natural joy, or bliss. After all the first order of dignity is the defense of one's life; the second, the defense of one's nature and being—one's spiritual-ontological identity. The two merge as one in this first war against extinction that we are called upon both by living nature, and by human nature and being to wage.

If we were to lose, our 'successors', unviable, alienized, cyborgized humanoids, *Homo alienus*, would record the war in their way. Winning, we will record that coming to the point of transmutation that acquires the element of self-destruction, we diagnosed the pathology and fell back on our primordial beginnings, foundations, and traditions for healing and nourishment; that we rediscovered natural, authentic, human selflifeworld, and authentic, human nurtureculture. We can still someday say that humankind, pushed against the sci-technic wall, pushed into the malnourishing, urban, reservations/zoos of urban sprawl and ghetto, pushed into exploding socioculture and human reality followed by human electronification—the electron bombing and explosion of the human spirit and soul; that finally we received our human revelation, awoke from our nightmare, and re-embraced our authentic nature, being, identity and reality, as persons, organic, families, communities, and as a human species and subspecies.

## 23. Intershining Human and Sacred Light

The light of human authenticism predates the light of religion. It is the light of our original magnificense and natural purity. It emerges from the deep primordial and reveals our original, authentic, human self, consciousness and lifeworld. Its bursting upon the West is the climax of the probing of romanticism into the human primordial pelvis. The seminal seed released, found the egg released by renaturalized consciousness via anthropology, sociobiology, ecology, ethology, rediscovery, revelation, and other agents of the last human spring, and reconceived an ancient realism—the revelation of natural, authentic, human reality, nature, being and spirit; their selflifeworld and nurtureculture. This deliverance can complement or intertwine with religious redemptions. Without necessarily opposing them, it intershines with the great religious truths of the various 'civilizations' of humanity with the complete *human* light of humanity. The former substantially redeem their peoples in their time and place. Authenticism's human revelation redeems humankind, the species, by redeeming the West which leads humankind toward pending extinction or decimation.

Just as the planet Earth is now seen through the ecocrisis as the homeland of all of humankind, though our home has always been the Earth—specific regions of it for specific groups; so also, the condition of human authenticism—natural, human selflifeworld with natural nurtureculture—can be again the home of the mind, spirit, nature and being of humankind. Home is where one goes for the basic matters of life; to eat, sleep, share, grow and to love, embrace and defend spirit and soul; and to defend one's selflifeworld; to make a last stand. Return home, then, 'civilization' peoples; to the real thing, to your human authenticism—to your authentic selflifeworld. Long gone away, separated and adulterated, you, now at long last, *can* commence the real and good journey; you *can* go homeward again. A movement of recovery and preservation is now our desperate duty to over four million years of our evolutional, human creation.

# CHAPTER III

# THE REVELATION: NURTURECULTURE AND NURTUROME

## A. Short Forward

IN NATURAL, AUTHENTIC, human society, parents, family and community are prepared to receive living nature's genetic bequest into the arms of her cultural bequest upon the birth of each child; for, they have been faithful to living nature by bequeathing, from generation to generation, the natural nurture and culture–nurtureculture and nurturome–of their species triecologic self and selflifeworld within Earthlifeworld. Genes–biology–upon birth find their home in both the physical body and the social-spiritual-ontological 'body' of the family-community's natural nurture and culture: two natural bodies (realms) coevolved within Earth's organismicity of life. Living nature's genes and genome are ignited with the breath of Earth-air, water, phenomena of ecophere, kindled with the nurture-love of mother; and the flame of our human spirit and being is tended by each generation's natural nurturome–of organic family-community and Earth-ecosystems: this is the passage of human being through time.

The hope for 'civilization's partial redemption through renurturalization (renaturalization) lies in a deep, human revelation of natural human being and spirit. Just as the person who is chronically ill or handicapped must understand and accept the nature of his limited human condition; so it is that 'civilization' humanity must realize and accept that it is infected with the alienism of 'civilization', must treat itself for that illness back into significant healing moving toward human growth-fulfillment and perpetuation rather than continue toward fantasies, illusions and decimation. Only when we realize we are fallen–have lost much of our natural human being–can we, then, seek and recover some lost elements of our lost spirit and being and work our way back into our role in the greatest and longest-running play we can join, evolved Earthlife, produced and directed by Earthlifeworld (ecosphere).

# B. Slaughter of Great Myths

## 1. Glimpse of Human Authenticism

As a consequence of the revelation of human authenticism, the concept of 'the authentic' as that which is original, genuine and/or real, is expanded beyond its usual application, which has been to a limited number of objects and activities authentic and characteristic to a distinct person or cultural group. Authenticity (as opposed to existentialism's use of the term and for which reason the term will be avoided) is expanded to all objects-phenomena of the whole human realm, i.e., to everything about 50,000 year old modern humanity, its human being, reality and selflifeworld. This volume affirms the revelation that there is a condition or state of humanity that is authentically ours, which is termed, herein, *human authenticism*. Within this condition lies our original, authentic nature, spirit, being, identity, nature, and reality that are fitting and belonging to us, that are inherent in us through Earth-life's–living nature's–primordial, evolutionary ecological design and creation of us.

Our human authenticism (of our nurturome) has existed for

about 300,000 years, and (of our subspecies, modern humanity) for about 50,000 years. For, at least, 5,000 years, 'civilization' has experienced expansions of domination and oppression alternated by contractions of substantially regained autonomy and freedom for the humanity ensnared or conquered by 'civilization'. Despite the relative 'success' of American 'civilization'–and partly because of it and the expansive, sparsely populated continent where it arose—, most of the readers' grandparents grew up more within Earth-folk lifeway and consciousness as opposed to urban, 'civilization' lifeway. Over eighty percent of our great grandparents were Earth-folk people of Earth-folk (socio)culture and consciousness. They remained unpenetrated by electricity, radio, secondary education, indoor plumbing, the automobile, television and other urban phenomena that would eventually adulterate and disintegrate organic, substantially autonomous, organic family and community. The a'civilization', Earth-folk paradigm of socioculture still largely prevailed over urban, industrial socioculture. Most Americans and perhaps Westerners in general were substantially within a substantially natural, authentic lifeway and conciousness.

By way of further elaboration, the concept and philosophy of human authenticism moves beyond particular, authentic phenomena of ethnic (socio)cultures, such as authentic (Mexican) food, authentic (German) Earth-folk dance, authentic (Aztec) pottery; and beyond actions or traits of a person pertaining to the "authenticity" of one's life and character advocated in much of existentialist philosophy. We move to contemplate and speak of all natural, *authentic human* elements-phenomena that are in line with our natural, authentic patterns and paradigms and that participate in organismic, human being and selflifeworld within Earthlifeworld. We contemplate, then, our primary authentics as Homo sapiens sapiens: human nature, spirit, being, reality, whole self, family, community, lifeworld, nurtureculture, whole mind, values, beliefs, perception, intelligence, mythology, knowledge, nurture, goodness, beauty, childhood, will, truth, reality, freedom, purpose, language, understanding, 'technology', play, work, love, and ecoregion, etc. (*See* "The Revelation of Selflifeworld," Part B, below.) We move, then, to reveal and describe everything that constitutes our natural subspecies of modern humankind and its natural sociocultures (societies).

Human authenticism is the growth-fulfillment and perpetuation of the human nature, being, identity, spirit, reality, nurtureculture and the selflifeworld (structure-function-dwelling) of these, these all being natural and authentic to our human species. Human authenticism, the philosophy, reveals this human megaphenomenon–this authentic human condition–of human authenticism, which could also be termed, human naturalism, (distinct from other naturalisms). It is the breakthrough into the vision and revelation of human authenticism that yields this new, complete concept of 'the authentic' to 'civilization' consciousness, providing a vision of original, unadulterated, unfragmented, prehistoric consciousness, nurtureculture, nurturome, selflifeworld, and lifeway. This vision, understandibly, both frightens and astonishes the Western mind with the conceptual restructuring of philosophy, understanding, human reality, human nature and being. This restructuring is the rising of the authentic to unite with and reconstitute the true, the good, the beautiful and the real, forming the ultimate test of everything; forming the long-sought, complete and unified understanding and philosophy. Ultimately, this is a consciousness that science cannot achieve nor effectively promote (as explained later), since science achieves its own alienized consciousness. Authenticism holds that there is a natural, authentic, human consciousness, belonging to/with our species; that it unfolds freely and naturally within its own medium, its own natural, mental, social, spiritual-ontological, material, phenomenal, human nurturome authentic to us–within its own, natural selflifeworld. This selflifeworld, its nurtureculture, and consciousness are substantially diminished, artificialized, adulterated and transmuted by 'civilization'. (*See* nurtureculture below.) The degree of our natural and authentic growth-fulfillment-perpetuation–our physical, mental and spiritual nourishment, from birth through maturity–is, thereby, the degree of our natural, authentic, human consciousness and being. Our natural consciousness is of authentic, human nature, spirit, identity, being, reality and nurtureculture; and, is purely natural, as a blend of two naturals, of two natures. The first, innate, genetic programming or instruction, manifesting such as, birth, suckling, walking, talking—all behavior, function and characteristic arising from genetic instruction; the second, natural nurtureculture, our nurturome–everything after genes that provides for species' social and personal growth-fulfillment and perpetuation. (*See* "The Root Human Birthright," Part 3 of Ch. II.) These two natures are intergrown; the second growing out from the first. For, though it is amazing and repulsive to 'civilized' humanity and to the West in particular, its oldest, most central and paramount myth must now be abandoned for the sake of human preservation. This is the belief that (a) nature and (b) nurture and culture are, by definition, two distinct and separate phenomena: the belief that human (genetic) nature, is of living nature; 'culture' and nurture are of humanity. The break from this conceptual shackle may start for some with the knowledge that simple, human culture and rudimentary, primate culture existed before the artificial 'culture' of 'civilization' that is so conspiciously in conflict with living nature and her human nature. (*See* "The Revelation of Nurtureculture and Nurturome," Part 10 below.)

## 2. Modern Myths Vs. the Human Journey

Great myths die hard and with staunch, hostile resistance. But they die, nevertheless, when they clash with the truths and realities of life itself. Life buries on its pathway all myths that threaten the journey itself. And, evolved human life, nature, and their lifebeing-forces have buried or decimated, downsized, dispersed, and rectified, all societies and 'civilizations' when their untruth becomes a threatening burden to the Human Sojourner, and the Sojourner, Earthlife.

Within, sci-technic humankind, some of us having experienced substantial, renaturalized consciousness simply have rediscovered too much about ourselves and our historic and prehistoric ancestors from anthropology, biology, sociobiology, zoology, ethology, ecology, and related fields involved in rediscovery of primordial, authentic, human nature and being, not to realize that just as every other species has its natural environment and every other social, primate species has its natural, physical and social environment, and simple or rudimentary 'culture'–nurtureculure; likewise, humanity is now exposed by the light of human science and respringing, naturalistic philosophy as being a part of living nature, not the different and separate ruler of it. The myth that man has his own nature that is beyond living nature and creates his own cultural nature and being that is opposed to, outside of, and/or above living nature; this myth is now revealed as a statement about adulterated, artificialized, alienized, sci-technic humanity only, i.e., it overlooks and denies the fact that our original, authentic, human nature is still present and asserting itself whenever able. This myth makes impossible the very idea of natural, human culture–nurturome–that is of a natural species; this myth, my dear readers, my fellow humans, can no longer be held. For, now, at last, it threatens the Human Sojourner.

This myth is dying as much by the barrage of research findings in the human sciences that have recently struck it, as by simple observation of newly discovered Stone Age peoples that occurred in and around the decade of the seventies. There is nothing new in the known existence of Stone Age peoples; but, these were not only still pristine–untouched by 'civilization'–they were discovered in the milieu of quickening re-evaluations about pre-'civilization' humans that were gathering force in cultural anthropology, ethology, sociobiology, transpersonal pyschology, ecology, and other realms of thought, disicplined and undiciplined.

## 3. The 'Savage' Myth vs. Human Science

In our time, a growing number of thinkers about modern humankind and society are being forced to accept understandings that 'civilization' thinking, and especially modern scientific and philosophical thinking, have previously held as given unthinkables. These new understandings include: (a) the age of our subpecies, modern man, now held to be about 50,000 years, instead of about 15,000 years; (b) the new understanding of the true and authentic living patterns of pre-'civilization', Earth-folk socioculture; and (c) the new understandings of the portion of our species still within a'civilization', Earth-folk socioculture, still significantly or largely free from the fragmentation, surrogation, diminishment, disintegration and adulteration from 'civilization's penetration of it. Such new understandings reveal a competent, loving, civil, cooperative, egalitarian humanity outside of 'civilization'; they reveal folk, a'civiliation' communities of peoples with long life-spans, largely disease free through their lack of urban stress and lack of contact with diverse gene pools that harbor diseases beyond their immunological system. Further, it is now understood through the merging diverse fields of thought that this non'civilization' humanity was and is in general ecological balance with the animal and plant worlds. Taken all together and absorbed, such realizations are manifesting a threshold revelation of non-'civilization' humans: that they were/are not savages and/or of inferior intelligence to us after all; that they were/are merely of more simple and natural socioculture. And that the quality of this humanity and its selflifeworld, by any unbiased measure, approximates, to say the least, us and our 'civilization' selflifeworld, with the advantages and disadvantages of the two balancing out *roughly* the same; and, perhaps, as some claim, the former may be superior to the latter. This threshold revelation ripens the human mind for still another, a final and complete revelation of conceptual deliverance, the revelation of human authenticism.

Afterall, such an unbiased view leaves any general preference for 'civilization' grounded in subjective, personal and cultural bias— it is what we are accustomed to and comfortable with, what we are born and raised within. More important, this 'objective' evaluation of the two basic human sociocultures (*See* Natural, Authentic Human Society vs. 'Civilization' below) crashes open the gates for the selection of values, elements and lifestyle modes that are outside what is enculturated and accustomed in us. These new elements and values found to be more suitable than conventional ones to our authentic, natural being, identity, and reality, reap a two-fold accomplishment: they expand and legitimize pluralism in our enculturated 'civilization' society; and, more important, they open 'civilization' man and lifeworld to values that were previously unattainable and taboo, those lost and missing values of both pre'civilization' and a'civilization' Earth-folk humanity, and their natural, organic family, organic community and selflifeworld. This amounts to a new romanticism sprouting in new, rather firm scientific and philisophical soil, and a neo-romanticism without the doctrine of individualism, which has been to generally pushed into a doctrine of egoism and greed. It is a neo-romanticism approaching a new, final realism, approaching an authentic revelation of human redemption.

## 4. Threshold to the Human Revelation

The above, then, at last opens a clearing, the threshold to human authenticism. At least a few Western intellectuals in the past have entered upon this threshold; and, it is perhaps a fair estimate that, since the middle of the seventies, several have pushed well into it, into the light and warmth that precedes our next and final resurging and respringing of human authenticism, the respringing of our species' spirit, being and reality.

All light meets the objects that it enlightens. Authenticism exposes humanity before and after adulteration by artificialized, alienized 'civilization'. It exposes sci-technic man as unnatural, as having largely renounced kinship and allegiance to living nature, human nature and being, and as having violated the integrity and laws of both. Authenticism's light reveals, at the same time, the redeeming truth, that man is by nature a part of living nature, and has, therefore, been violating himself with adulterants of 'civilization'. Human authenticism reveals the 'culture' and the quasi selflifeworld of this unnatural, estranged, adulterated human. We are capable of seeing either illusions or reality. The new light makes reality discernable to those open to it. To all those who sense and will seek a long, lost, authentic self, lifeworld and authentic, human nurtureculture, the light exposes 'civilization' socioculture as destructive of authentic, human endeavor, values, spirit, caring, will, purpose, consciousness, and other characteristically human elements-phenomena. This deleterious socioculture now exposed to those who can and will see, is part of the evidence for a natural, human (socio)culture authentic to modern man. For, upon realizing that the modern selflifeworld is largely and increasingly unfitting and unreal, one will if one dares, ask *What was the real, authentic, human selflifeworld like, before this?*. Authenticism makes it possible to dare, by providing, at once, the beautiful, true and good answer. It describes a real selflifeworld that was and is natural and authentic to us through its perpetuating structure-function-dwelling–through its dual nature and being: natural biology and natural nurtureculture. (*See* nurtureculture in Part 10 of this chapter.)

The real selflifeworld and reality is for modern man, Homo sapiens, the same as it was, in rough, general principle and essence, for our predecessor subspecies, Neanderthal man, and, further, as it was for all present primate species living within any nurtureculture, however rudimentary or simple: it is authentic, species, genetic living nature functioning intergrown and interfunctional with the natural, authentic nurtureculture of each species.

## 5. Nature Vs. Nurture and Culture: the Key Myth

It is natural, human culture that has been the main, key, illusive, missing value and concept in man's quest of self, understanding and fulfillment and a philosophy for these. Authenticism reveals this as a *natural nurtureculture* (*See* "The Revelation of Nurtureculture", Part 10.) authentic to our 50,000 year old species, modern man. Thought about human nature, heretofore, has conceived and granted all the other naturals except natural culture: natural biology, food, clothing, shelter, leisure, work, instinct, senses, behavior patterns, character and personality traits, et al. But, these have not been pulled together into attaining an accepted, definitional conception of a purely, natural, human being worthy of emulation, identification or embracement. A natural, human culture has been inconceivable. In traditional thought one illusive answer to an elusive question has slipped by the thinkers about the naturals and human nature: the answer to the question, *If there is a natural man, what would be his natural social life and culture?*. The question could not be asked because man without 'civilization' was believed to be a hostile loaner incapable of social or community behavior. The entire written history of mankind, after all, has depicted the general idea that humans progress *with* historic 'civilization', with some setbacks, from the subhuman savages or barbarians of prehistory to a civil citizen of the city or state. How contrary and ironic that the Christian Bible of the West, says that after Adam and Eve, asserted as the first humans, misbehaved, they were shortly outdone by Cain, a son, with the slaying of his brother, Able; a Cain who, under God's permanent condemnation, built the first city, a hallmark and the major phenomenon of 'civilization'. In Western history books the Germanic tribes, still composed largely of Earth-folk, a 'civilization' villagers, after retaking ancient pastoral lands conquered by the 'civilized' Romans, were labeled the *barbarians*.

## 6. The Fall of the "Savage" and Wilderness Man

This a priori notion and given view of pre-'civilization' humans as savages has been an unpenetrable barrier blocking the conception of an admirable or noble 'natural man' somewhere in our past, hovering within our personal and collective self, longing to rise from suppression, from the prison of 'civilization' socioculture. Rousseau, though not the first, was the most eminent of the rebellious, Western thinkers to break through this barrier without the ability to actually break it down. Some in the European masses anticipated the spirit of Rousseau and seized as conceptual treasure, the shining accounts of the New World Indians as egalitarians of noble character given by some explorers and colonists. The light at the center of Rousseau's famous "illumination", while walking a country road, was the realization that we were preceded, not by a brut savage, but by a 'noble savage', a "natural man"; a natural human whose fulfillment and freedom become thwarted and oppressed by 'civilization' disguising its oppression and greed with grandious acheivements that benefit and please a wealthy, power-corrupted, royal class, and beguile the middle and lower classes. Rousseau gets much of the credit for spreading in the popular mind this competing image of what our prehistoric predecessors were really like. Still, this key idea was never to take deep root, while many swirling out from and around it, became the lasting legacy of romanticism.

A central problem was and has been terminology. The term, 'noble savage', was first used in Dryden's play, *Conquest of Canada* (published 1672), though the concept can be traced to ancient Greece, where Homer, Pliney and Xenophon idealized the Arcadians and other primitive groups. Later, Virgil and other Roman writers gave comparable treatment to the Scythians. Whatever number of noble or admirable characteristics one may attribute to or find in a 'savage', the fact remains that, a general condition of human savagery, even if complimented by a newly discerned nobility, cannot be made readily and thoroughly acceptable as the generally good and exemplary human condition. 'Noble savage' entails contradiction. Perceived as living in humanities "natural state", why not call this human, "natural man" as Rousseau preferred. The answer no doubt stems largely from the colorful, primordial, controversial nature of the term *noble savage*, and even more from the large opposition–for politcal and metaphysical motives–to the American Indian being given acceptance as an admirable race of people, or even as human beings. The victory of this ambiguous term denoting both admiration and derogation throughout the eighteenth and nineteenth centuries foreshadowed the conquest over the peoples it referred to by the influx of Europeans and their socioculture, conquest to the point of their ongoing decimation.

Although Rousseau was out in front of his contemporary, leading thinkers that admired the 'noble savage', he was blocked from carrying his revelation of a "natural man" constrained from fulfillment by the artificiality of 'civilization' socioculture, into the even more revelatory concept of a natural humanity possessing a natural family and community. For, Rousseau–having himself grown up with very little family and community, and being a misfit or outcast much of his life–accepted the contemporaneous idea that natural man was a loner with only occasional social contacts. Rousseau, after correctly placing his natural man outside the oppression of 'civilization', experienced his main and most disasterous failure: he, alas, placed his natural man alone in the wilderness, unnurtured by–non-interfunctional and non-intergrown with–natural family and natural community. The very same anomalous and original upbringing and education that contributed to his original thought and genius also blocked him from completing his revelations and his system of philosophy. Add to this, Rousseau's attachment to books, music and writing, key characteristics of 'civilization', Rousseau was put into an irreconcilable conflict. Such conflicts often enter partial awareness, but usually are blocked from full awareness for the sake of self-esteem, well being and sanity. The well known ethnocentricities that block the scholar from a full discovery of objective human truth and reality; these, Rousseau had substantially broken through but was unable to break free and clear from in his presented philosophy. To complete his revelation and philosophy untimately required that he reject nearly all his own life, self and world. The magnificent rebel fell short of the necessary conceptual self-sacrifice.

The other two conceptions perhaps coming closest to a natural man have been pastoral man and wilderness man. Pastoral (rural) man is ruled out as one who is intermediate, or on his way from wilderness man to urban man; not to mention the traditional, propagandized condescension dealt rural people by educated urbanites. This leaves us wilderness man, since those looking for a natural man, do so from their dissatisfaction with or renunciation of urban man. We arrive, then, back at wilderness man–the 'noble savage' of the 18$^{th}$ and 19$^{th}$ centuries, a long-standing myth and now understood by anthropology as a contradiction. For, man is a social animal; in order to be human, he must have family and community. Any community of people or extended family (which is family attained to basic community), living in the wilderness, are living in a social community, a human, social structure-function that is in harmony with, but not *of* the wilderness, not alone and created *by* the wilderness itself. Anthropology now has recently dispelled the myth of wilderness man by tracing the nomadic hunter-gatherer band that travelled, with family-community and nurtureculture, its familiar seasonal rounds, back at least to early Neanderthal; and, in fact, present primate knowledge even reveals simple primate community substantially like prehistoric Earth-folk community.

## 7. Recent Quickenings of The Last Human Spring

Anthropology, sociobioligy and ethology, (to mention just three of the fields), during the seventies and eighties, while providing for the destruction of wilderness man, more importantly, uncovered two truly revolutionary things about our subspecies, modern man, already mentioned, i.e., that he is much older than we had thought–actually around 50,000 years old; and, that his 50,000 year old nature has generally been cooperative and community oriented. In other words, the fall of wilderness man was accompanied by the fall of the idea that 'cooperative,' 'amiable' modern man emerged recently and rather rapidly from wilderness, 'savage' man, and won out in a competition or struggle with the Neanderthal 'brut savage' that surrounded him. In short, and to be blunt, virtually none of our traditional ideas about pre-'civilization' man have survived the myth slaughter of recent anthropology, sociobiology and ethology. We have been presented some new and rather firm, key ideas that amount to redefinition of our immediate, ancestral, prehistoric humans, and of Neanderthal man, who emerged about 300,000 years ago. These new understandings and those of human ecosystems, sociobiology, ethology, and some key others, old and new, are carried further and swept up in this volume (and in some degree by some other thinkers and writers known and unknown to this author) into a revelation of redeeming awareness, the revelation of human authenticism.

With wilderness man dead, and pastoral man alive (and mostly well) but disqualified as wholly natural through his being partially denatured; from whence authenticism's natural man? Here, then, is an opening, initial argument for natural, authentic man. Authenticism asserts, and asserts the undeniability of the following assertions: That, whereas modern man is roughly 50,000 years old, and 'civilization' began arising only 5,000 years ago; and, whereas 'civilization' as a human lifeway did not spread to encompass a majority of modern humankind until some time during late 20th Century (*See* Chapter VIII, "NATURAL, AUTHENTIC, HUMAN SOCIETY VS. ALIENISM".); it follows that the phenomenon of 'civilization' lifeway cannot be natural to the subspecies, modern man; but, rather is artificial and alien to him. (It is crucial to be aware that most people living within any 'civilization's' territory have not traditionally been living within its lifeway until they are living in a city of more than 5,000 or 10,000 inhabitants; and, this definitional criteria is rough and excludes those 20th Century small towns that have been penetrated and largely transmuted by the communications and technological revolutions–the precise moment of switch being that of nurtureculture mutation–the definition and theory for which is given later in this chapter and elsewhere in this volume.) This conclusion holds from the fact that whatever is natural to a species is continuously present in the species with two exceptions: one, during species' experimentation when elements-phenomena outside its nature are being tested for incorporation; and, two, whenever a species is in species transformation from one species to another. 'Civilization' lifeway, then, with its increasing, alien, artificial elements-phenomena, is either being tested as a new, mutant, sociolcultural, element and phenomenon for possible incorporation into and a feature of the species nature of modern man; or, it is actually in the process of being incorporated as new element within the process of species transformation of modern man into a new human species. Authenticism holds that the former of the two possibilities is the one which is occurring in modern man with the phenomenon of emergent 'civilization'. 'Civilization' can only be tested and, as with past 'civilization's, repeatedly fail the test, aflicting anomoly, deterioration and injustice on persons, families, communities and societies.

Authenticism holds this following its assertion that 'civilization' is not merely a new, alien, species element-phenomenon outside of the modern human species' nature, but is, also, alien, by virtue of its artificiality. And, since the artificial is the very opposite of living nature, and human nature and being, it cannot be successfully incorporated as new, natural element or feature of natural, human species' nature and species' being. Such incorporation would require the invention of artificial genes and nucleic acid capable of being accepted for surrogacy by these naturals, which have perhaps a billion years of self knowledge and identity, which effects a general rejection of artificial, alien elements. 'Civilization', therefore, is predestined as alien, artificial elements-phenomena, to continuing failure of the test for inclusion into the organismicity of natural humanity and its selflifeworld.

'Civilization', then, has not, actually, been incorporated into the species human nature and being of modern man. And it cannot be so incorporated, according to the theory of alienism presented later herein. 'Civilization's' rejection by human nature and being has been substantially unconscious; some exceptions are the hermit, the recluse, the sociopath, migrations back to the rural setting of one's youth and rejection of urban life by many rural people in many cultures of many centuries. And rejection is necessary, proper and inevitable to the survival of human nature and being. The *seeming* acceptance by humanity of 'civilization' lifeway, wherever it arises and opposes natural, authentic human a'civilization' Earth-folk lifeway, is, in essence, an illusion and human delusion. As explained in Chapter V, NATURAL, AUTHENTIC, HUMAN MIND, it is only the intellect, rebelling and parting from the whole mind, that accepts any artificial, alien elements into its room in the household of the mind. The acceptance is by the intellect, which at the moment of nurturecultural transmutation, seizes the lead of the mind. For the whole mind, this is subordination both to the intellect's new power and the presence of the alien artificials. The mind's natural, balanced perception is overthrown; egoself emerges (*See* Ch. IV, ORIGINS AND ARISINGS OF EGOSELF.), beholding life and world separated, surrogated, disintegrated and adulterated into quasi life and world. (*See* Selflifeworld below.) The resulting quasi 'subspecies' breed or humanoid is not of natural, authentic humanity, from natural selection and mutation; but a quasi 'subspecies' of synthesis, a quasi human blended from unnatural, alien penetration of our natural species' selflifeworld by *alien*, artificials effecting our transmutation into Homo alienus. (*See* Ch. V, NATURAL, AUTHENTIC, HUMAN MIND.)

## 8. Rediscovering and Re-embracing Human Nature

The existentialist view that there is no human nature is put succinctly by Merleau-Ponty, a 20th century French phenomenologist who concluded that "it is the nature of man not to have a nature". The common observation that leads to this common, unsound conclusion is the great diversity of 'cultures' existent now and in the past. This common conclusion is false because, upon study of ancient 'civilizations', this diversity shrinks from great to substantial variety; and, upon study of 'pre-civilization' societies, e.g., Neolithic (circa 10,000 to 5,000 *Before Present*) and Paleolithic (circa 300,000 to 10,000 B.P.), we find the great diversity has vanished. Unbiased, clear study reveals the essential objects and behavior characteristic of and authentic to our species' human nature and being, some of which is still intact, though inundated progressively by unnatural objects, behavior and phenomena. The secular idea of humankind being outside of living nature is generally confined to ego-intellectuals living in socio-cultural systems of the last 150 years or so–societies that are highly artificial–and, to those rejecting the core idea of romanticism: a primordial, natural humanness in our past, partially retained through the simplicity and tradition of rural societies, and buried within us by alienized, artificialized, urban societies. Various religious ideologies also place man outside living nature by affixing him a dual nature, of divinity and deviltry. This human separateness idea, this doctrinal, nature-humanity apartheid–that humans are separate from living nature–is egocentric, ethnocentric, industrial society centric and religiocentric, and rests upon a convenient, purposeful ignorance and/or distortion of human history, pre-history, ecology, human ecology, and ethology. This false, predominating, nature-humanity apartheid doctrine of existentialism and Western religions, corrected and truthfully stated by human authenticism becomes: it is the *condition* of *'civilization' man* not to have much of a nature left, after having his authentic, natural, species nature adulterated, artificialized, denatured, and transmuted by 'civilization's' unnatural (alien and artificial) elements-phenomena.

Since the 18th century the question of the place of man in the world–Earthlifeworld–or universe has been formulated in a new, non-biological way: man as a cultural being in opposition to the natural realm. There have been several, specific forms of this dichotomy, which authenticism reveals as a false dichotomy. The main forms are as follows: man is the only being who is self-conscious, who has a language, who uses symbols, who employs tools, who freely plays, who possesses a history or possesses the distinctive kind of intelligence for these. Over and against this modern view, i.e., that existentialist man makes himself, is the "archaic" view that man was formed by the gods. His history is given in the myths of the primordial establishment of things and his solemn responsibility, along with every other living thing is to fit himself within this given world. This "archaic" view either gains strength or is modernized under two presumptive conditions: (a) that the gods were/are parts of living nature; and (b) that, in some theisms, God creates through nature, after first creating living nature.

This older view–basically that humans are a part of creation–has recently regained strength in Western 'civilization', ironically, from science–ecology and human ecology–which is rediscovering that contemporary man, also, has a solemn responsibility like other living things to fit himself, by a necessity for welfare and survival, within the given, natural world. The shattering truth now increasingly revealed in the light of ecology, cultural anthropology, ethology and sociobiology is that man is a natural being after all: that 'civilizations' clever sociocultural systems count for naught now that *Chicken Little's* cry, "The sky is falling, the sky is falling!" has come to reality: it is sublime that our destruction of the ozone layer fulfills the funny proposition of a story meant to reassure children of nature's stability. In fact, several threats to our survival through the destruction of essential parts of the natural world by emerging parts of our cultural (artificial) world beam brightly into the asphalt jungles the same "end of the world" message. And, combined with the message of anthropology that man did evolve from non-man–from ape–the realization is growing and pressing hard against modern, existentialist thought that man is, by natural nature, a natural being primarily, to say the least, and a artificial, enculturated one only secondarily. It is only the rationale of the intellect-egoself with its vested interest that can hold a contrary, resistance view now: for, even this rationale can at best, now tenably propose a compromise nature, a dual human nature–half natural, half artificial enculturated. This has been come to be the best defensive concept usable by the Western, secular intellect to attempt to block the spreading human revelation of the pure nature of the human species. However, it also is a desparate grasp; for, it is revealed as falsehood and illusion at its core, by the revelation of human authenticism.

## 9. The Fall of 'Culture' as Human Distinctive

Despite recent requickenings of our human spirit and being, the above, key obstacle and misconception, aided by others, still blocks a fuller, wider resurging of our human spring. Our concept of human culture–of socioculture and society–in general holds these to be in opposition. It is this dichotomy– the essence of human nature and humanity conceived to be outside and opposed to living nature that is false and illusional. In truth, it is only the artificial 'culture' of 'civilization'–and people artificialized by it–that opposes living nature. The key, central, magnificent, redeeming revelation of human authenticism is that man has a *natural culture*; modern humanity itself, Homo sapiens, has a natural culture and socioculture authentic to it as a species of humanity; and, our preceding species (Neanderthal, Homo erectus, and others) have had a natural culture that is authentic to each of these species in the family of man. Of course, these cultural natures of different human species indeed overlap in their evolution along with their general, species natures. Natural culture has been termed nurtureculture and *nurturome* in the philosophy, herein, to separate it from what we have been thinking of and designating with the term *culture*. The difference is that nurtureculture, unlike 'culture', is post birth, evolved nature– nurture and socioculture–evolved within genes because they actualize and fulfill the genetic nature of the species:

nurtureculture is fully and correctly in line with evolved, genetic nature; artificial 'culture' of 'civilization' is out of line. Nurtureculture is the actual growth-fulfillment and perpetuation of the 2nd of the two natures of all highly social species: (a) biologic nature and (b) social, nurturecultural nature; it nurtures from birth to death and perpetuates through the unfolding growth of personal and species being, nature, identity, reality and spirit. 'Culture', by contrast, is all object, phenomenon and behavior that does not do this; but, instead is deleterious to—restricts, obstructs, limits, diverts and transmutes—this unfolding growth-fulfillment-perpetuation of human nature, being, spirit and reality, with their selflifeworld within Earthlifeworld.

The revelation is that genes (and genome) cannot carry evolving life without nurtureculture (and nurturome). The newborn child requires nurtureculture to achieve growth-fulfillment-perpetuation (reproduction). And, these three are not just genes and biology, they are family and community structure and function—social relationships—past on with genes in a union of these two elements of evolutionary life. Families and communities and species die out not only from unsuccessful genetic features; they die out from unsuccessful family-community, social, and cultural features and behavior.

## 10. The Revelation of Nurtureculture and Nurturome

The basic revelation of natural, human nurtureculture and nurturome has four parts: (one) our human subspecies, Homo sapiens sapiens has a natural society and culture—a nurturome—authentic to our 50,000 year old subspecies, which evolved from our 300,000 year-old species, Homo sapiens, with its own natural (socio)culture—society; (two) *alien*, artificial *culture*, created within 'civilization' and increasingly predominant within sci-technic 'civilization', first penetrated, some 5,000 to 10,000 years ago, natural, authentic, human socioculture—nurtureculture, our nurturome–and progressively spread within humanity; (three) *artificial culture* of 'civilization' is both alien and decayous to natural, authentic, human socioculture–(society); while, natural nurtureculture is intergrown and interfunctual with species', inborn, biologic nature, and together comprise the two natures of natural, authentic humankind; (four) these two natures of modern humankind (and previous human species), (a) genetic nature (genome) and (b) sociocultural nature (nurturome), are in humans inherent–passed on from generation to generation–until the penetration of artificial, alien 'culture' partially replaces natural dual reproduction of these two natures with historical, quasi reproduction or misreproduction: when alien and accelerative change arises as anomalous, transgressive, and outside of slow, natural, adaptational and perpetuational change of human nature and being to the whole of living nature in which they are ecologically intergrown. Thus, alienism progressively decayes natural, human socioculture until the dissolution, decline, or the destruction and reconstruction of the particular, 'civilization' socioculture. The resulting adulteration of humanity produces a progressing unhuman, denatured alien being, which is prolife– can reproduce its biologic entity–but is anti-human being and spirit–cannot reproduce its nurturome's, socio-psycho-spiritual-ontological entity, its human soul. With his highly accelerated alienism, sci-technic man is progressively becoming one of the aliens for which some now search the skies and stars to find.

Certainly, we have abundant evidence that the pollution of nature ("the environment") by artificial elements produces disease and/or death in various species including man. Likewise, but more tragic, the progressive adulteration of human nature, being and unreality by artificial, alien unhuman, unnatural, alien being, and reality is the self-destructive, pathogenic delusion leading toward extinction or decimation of our human species. It is, however, halted and reversed by the revelation of human conceptual deliverance– the revelation of human authenticism. For, this supreme, magnifi-cent revelation to Western, secular, human consciousness reveals both our authentic, human being-force of growth-fulfillment and perpetuation, and the quasi, alien lifebeing-force of decay growing within our natural, authentic, human nature, being, selflifeworld, and life-being force.

## 11. The Fall of the Human-Animal Dichotomy

There are at least four ideas that destroy the dichotomy of man opposed to or distinct from living nature. First, the display by one or more species of humans of one or more behavioral characteristics not displayed by other animals does not resultingly remove that or those human species from the animal kingdom and natural realm. Many animal species display unique characteristics not found in their genus or family but remain in the animal kingdom via the remaining bulk of their organism's characteristics. Second, the so-called 'distinctive, oppositional characteristics' would have to have emerged from a source outside natural evolution; for, surely any species created by living nature is still a part of her—even though 'civilization' humans create and adulterate living nature and their own nature and being with artificials—living nature doesn't create artificials, only blended, synthetic 'civilization' humanity does. Third, the so-called 'distinctive, oppositional characteristic(s), e.g., tools, language, symbols, concepts or thoughts, all can be plausibly held to have evolved from forerunner characteristics possessed by one or more earlier ape-human species or ape species—to be a product and, therefore, a part of living nature. For example, regardless of when human speech attained to a substantial human vocabulary, it surely evolved from more simple and/or rudimentary speech of Neanderthal and Homo erectus—from speech precursory to Neolithic and 'civilization' speech; all evolved in rough principle from something similar to what has been recorded among present day chimpanzees. And, moreover, regardless of when the first ape species transformed a part of its intelligence from using sticks to obtain a mouthful of ants to sharpening and shaping sticks and stones for cutting, piercing and chopping; this change in intelligence, even if judged an increase, may be used to mark a point when ape animals became human animals or part of each (ape men); but, by no means can a change or step up in intelligence be defined as a step out of or up from living nature or its animal kingdom. Fourth, the main principle binding prehumans to humans is this: an animal species using naturally evolved intelligence to fashion tools to alter stone, wood, bone—natural materials—toward more effecient hunting, gathering, preparing, storing, building (shelter), and making clothes—toward natural human-animal endeavors fitting in with natural human purposes of living nature, *which includes purposes of human nature and being*, is still within the natural, animal and human-animal realm. Thus, our species of humans using spear, bow and arrow, utensils of wood and stone to secure and prepare food, the same wood and stone to build shelter, the skin from game to secure body warmth; all such are within the evolved nature intergrown living nature and human nature and being. To reiterate with an example: our species' use of fashioned oral sounds—words—to communicate naturally evolved objects, activities, ideas, thoughts and emotions that achieve through this means, natural ends and purposes of our species' human nature, being, identity, reality and spirit is within the natural, human animal realm. (*See* Authenticism and Language in Ch. X.) All of the above mentioned characteristics can no longer be used to separate humans from animals; for, they occurred in simpler forms in nonhuman primates.

In summary, modern man, the 50,000 year old human sub-species, is a natural being until artificial, 'civilization' element-phenomenon penetrates his being and nature and progressively adulterates his pure nature and being with artificial, alien con-stituents. The mere creation by human intelligence of something which imitates an essence found in living nature (as clothes imi-tate animal fur, spears imitate sharp sticks, mud or wooden huts imitate other animal shelters, e.g., branch and brush thickets, burrows and caves of mammals and primates) does not accomplish the elevation or removal of human intelligence out of animal and natural intelligence: it is still human animal intelligence.

'Civilization' man, in contrast, emerging about 5,000 years

ago and increasing into a majority of our species only sometime in our time, is distinctly a blend of natural, human-animal, being and artificial, quasi alien-human being—a blend of unhuman, denatured being, and natural being—an alienized creation, created through artificialization. (*See* Human Authenticism vs. Alienism, Ch. VIII.) It is only 'civilization' humans, not a 'civilization' Earth-folk humans of Earth-folk socioculture that have seriously diverged outside living nature and human nature; that have moved into a dichotomous position generally opposed to living nature, human nature and being. A 'civilization', Earth-folk society opposes living nature and human nature only incidentally and insignificantly rather than substantially or generally. General opposition is that which opposes the purposes and ends of the species—which are conducive to species basic purpose: the growth-fulfillment and perpetuation of its being, nature, identity and reality. In our 50,000 year old subspecies (and in, at least, immediate previous, human species) this entails a species selflifeworld and nurtureculture.

The test that distinguishes natural, authentic tools, techniques, behaviors, thoughts, imagination, speech, or other phenomena from artificial, alien ones is whether or not such means and their result-ing ends are conducive and fitting to our natural nurtureculture, selflifeworld and their natural, human nature, being, identity, re-ality and spirit authentic to our species.

It is not "We think; therefore, we are". But: we function/dwell within our natural, authentic nurtureculture and selflifeworld; therefore, we are *human* being, nature, identity, spirit and reality. We interrelate and share our being; therefore, we are.

## 12. Human Nature Rediscovers Home Within Living Nature

In summary, man is a natural being with a natural human nature and being; both human being and nature are intergrown and interfunctional with living nature: neither are or can intergrow with artificial, alien elements of 'civilization' socioculture, but only with those natural socioculture elements-phenomena that are still retained in the 'civilization' blend of natural constituents and alien, artificial ones. Artificial, alien elements of socioculture can only be intermixed, not intergrown or interfunctioned, in human being. That is, in the same manner that pollutants in food and water can only mix in the body accumulating in the kidney, brain, and other tissues decreasing the structure-function, likewise, socioculture adulterants decrease the structure-function-dwelling of natural human being, nature, spirit, reality, and their selflifeworld.

Reemphasizing, human authenticism reveals that our Homo sapiens' natural, authentic makeup is that of all natural being: not a being of artificial 'culture' nor of divine or supernatural essenses. Whatever divine elements of whatever divinity one elects to grant, within human nature, being and/or living nature itself; our so-called 'culture' remains, in fact, dichotomous: there is *our* original, authentic culture–nurtureculture–natural to our species and subspecies; and, there is the 'culture' deleterious to us, that of 'civilization', artificial and alien, which spreads within our natural, nurtureculture socioculture and within our being as pathogenesis. Again, that natural culture (nurtureculture) which is essential to our species in its natural structure-function-dwelling of selflifeworld–the tools, utensils, speech, relations, behavior, activity, et al., is natural (socio)culture, i.e., nurtureculture authentic to and belonging with our species. In contrast, that 'culture' that has been and continues penetrating human society gradually and progressively (with some cyclical setbacks) for some 5,000 years, reaching into and transmuting a majority of human lives only by sometime around the middle or latter 20$^{th}$ century, is artificial and alien to modern humankind. 'Culture' *as we have conceived it–'civilization' 'culture'–*is not the mark of humankind; it is merely the mark of 'civilization' humanity. *Authentic, modern humanity is only that portion not yet generally infected with 'civilization'; it has all natural being* having evolved through each species an accumulating, natural nurtureculture which was recently, 40,000 years into our 50,000 year old subspecies humanity, penetrated by artificial, alien 'cultural' elements-phenomena; the rest is history (of 'civilization').

A history, moreover, written by the majority infected with the decay which they must perforce perceive as good; for, it clearly does spread, has greater technical power, competitive, combative, conquesting and material presence. History ('civilization') acquires brute force, technical and material power, and materialism itself, all as both ethics and aestheticis. The shift, thereby, is made from natural, human socioculture's security and perpetuation to alien 'civilization's anxiety and wartime devastation–'civilization' is threatened from without by other 'civilizations' and states' brute force and materialism, and from within by the disintegration that is inherent in an entity infected with decay. Nearby natural sociocultures, henceforth, are threatened with destruction or social transmutation that are holocausts compared to the occasional, small skirmishes with adjacent nomadic or village societies, which, then, may or may not experience with pre'civilization' socioculture.

The social penetration is actually the collision of slowly evolved finite, human being–limited by/to its nature–natural, authentic, human society–human selflifeworld–with the phenomenon it cannot accommodate or absorb into it as intergrowth and interfunction, relate to peacefully (with trade or mutual respect) nor hold off through war: it is the collision of natural society with 'civilization', i.e., with city, state, nation or empire.

# C. The Revelation of Human Selflifeworld

## 13. Short Forward

The difference between living nature and 'civilization' is systemic, socio-psycho-spiritual-ontological, and antithetical. The 'civilizations' have never done justice to nature's human being and spirit. They inflate the egos of a few upon a bloody sacrifice of our, species human being. The flowerings of 'civilizations' are but brief weeds after a freak rain storm compared to human nature's woodland–our 300,000 year old Paleolithic nurtureculture. Still more proof beyond an intelligent doubt that 'civilization' is alien to living nature! Even more, 'civilization' lacks real beauty, while living nature is pure beauty free of adulteration, fragmentation, stunting, warping, surrogation, violation, waste and dysfunction. Living nature wastes nothing. 'Civilization' disguisingly violates living nature's soul. It will, as we cower, waste most or all of humankind.

'Civilization' itself is a bequeathed addiction to artificial, alien objects and phenomena that pollute and transmute natural human being into alien being, self, and lifeworld; it is the medium for the alienism destructive to humanity and all evolved, species beings of the Earth.

## 14. The Unified Field of Human Being and Reality

Human authenticism (the philosophy) recognizes three triecosystems: (1) the triecologic, whole self–bio-self, family-self and community-self, (2) triecologic selflifeworld–whole self, its life and world–and, (3) the triecology of bio-social self, human, selflifeworld, and Earthlifeworld. (Earthlifeworld has recently been termed by science *ecosphere* and *biosphere*.) These three ecologies, along with the concept of natural, authentic, human nurtureculture, provides for the unity of psychology and sociology; and these with psychoanalysis; the unity of social science and physical science; of natural science with human science. The unity of all the sciences is provided for in the absorption of science by the philosophy, human authenticism; a new creation that perhaps could be termed *philoscience* or *philosence*. (New, that is, to the intellect, not to natural, authentic, human understanding). It re-forms substantially what 'civilization' science has progressively disintegrated: forming a unified field theory of humankind's nature, identity, being, spirit and reality.

Precisely and theoretically speaking, however, authenticism expels modern science from human reality and being, as generally alien, idea system and phenomena. It supercedes study with natural revelation, returning to understanding, intelligence, and consciousness of predisintegrative humanity i.e., to tri-tri-triecologic humankind. Pragmatically speaking, it subordinates and domesticates science to a philosophy of human organismicity within Earth's organismicity. (*See* "Science," Part 29 of Ch. X.)

## 15. The Revelation of Triecologic, Whole Self

Evolving from some social rudiments in biologic nature, nature has evolved, with primate and especially human evolution, a second nature, i.e., a natural, social nature and being, of a natural social self–consisting of ego-self, family-self and community-self–intergrown with its natural lifeworld as selflifeworld. Before us, then, lies the greatest, grandest 'archaeological dig' of rediscovering human essences: to recover, beneath 'civilization's' overlays of phony, unnatural, 'socializing' egoselves and unnatural, social, lifeworld 'culture', the authentic nature and being of our species with its whole, triecologic self intergrown within our natural Earthlifeworld. Thus, we uncover and start the recovery of our natural, authentic structure-function-dwelling–selflifeworld.

Our 50,000 year old human subspecies and our much older species have come a long way on the journey into social and sociocultural nature. In predominantly biological animals, social life and being are largely limited to that which stems from instinct and senses–to that in the realm involving eating, survival and reproduction. These biological organisms are predominately separate and individual beings, interfunctional only by living nature's commandments: for cyclic, periodic reproduction; her commandment to brief intergrowth of mother and dependent offspring; and her commandment to groupings (e.g., herds, flocks) that facilitate this predominately biological being.

In our species of modern humankind (about 300,000 years old), the self is no longer an independent, mind-body subject; it has evolved into triecologic human self. The self has evolved beyond the confines of individual, biologic organism, identity, autonomy and being. Thus, human bio-self became–even long before ape-men–intergrown through living nature's evolutionary creation with family and community, as one of three intergrown, interfunctional portions of unified social organic being that create and constitute a natural, human, social self that is trinitious and triecologic. This triecologic self is intergrown in a second triecosystem: selflifeworld, which includes a natural environment.

Natural, authentic, human self, family and community are organismic, as an intergrown, interfunctional, unified, social being. They, intergrown to natural, human lifeworld, including natural environment, possess and fulfill a human nature, being, spirit, identity and reality; maintain and perpetuate (and reproduce) themselves and these. Whereas the laws of living nature hold that (1) organisms live and mingle in a general harmony with each other only preying on and competing with another organism when the law of survival commands it; and, (2) that all ecologies entail interactive, interfunctional balance: it follows that these laws of living nature apply to living nature's, human-animal, social self, and to its three portions–bio-self, family-self and community-self. These three portions have through the evolution of social life in the various, human species and prehuman, social, primate species, come to a triecologic social balance within each succeeding species. That is to say, in the natural, authentic, human social self– triecologic selfhood–the three intergrown portions of self–ego-self, family-self and community-self–obey both the law of ecological harmony–organismicity–and the law of survival. One part of the self does not abuse another part of the self because all self-parts are essential and vital parts of the whole human self; they pull together by natural, evolved system or law to maintain intergrown, unified whole self.

## 16. Whole Self and Triecologic Selflifeworld

In the unity of our social-psycho-spiritual-ontological essences, a magnificent form of natural being and spirit is present in our 300,000 year old human species, Homo sapiens. To pull apart this organismic human reality and being, in disection on the operating table of the artificialized, alienized intellect for the sake of science, is to slaughter humanity, a sacrificial massacre of ourselves. Sacrificial, intellectual surgery is for decaying humankind; unifying, human authenticism is for a humankind intent on rejoining Earthlifeworld's field of living beings. The basic laws of triecologic self pertain to triecologic selflifeworld as well. Self, life and world pull together to maintain intergrown, unified selflifeworld. In these two triecologies, it's all for one and one for all.

The three intergrown portions of triecologic self draw from each amongst them what is needed for their unseparate, triecologic autonomy, in a unified, trinitious autonomy of human being, nature, reality, and identity. The interflow after millennia (certainly over 300,000 years) is as natural as the interaction of organs of the body, forming a triecologic balance of the natural, human social system. This system, however, stops at organic community, excluding city and state–excluding artificial, alien culture and society. (*See* Chapter VIII, "Natural, Authentic, Human Society vs. Alienism.") Inherent, incidental, systemic frictions are provided for by inherent systemic lubrications and do not attain to dysfunctional conflict; e.g., sibling order and parental guidance lubricate sibling rivalry within family, family role in community lubricates family rivalry in the tribe, clan and/or village, as does allocation by consensus, custom, mediation and/or natural leadership of community tasks and activities. This triecologic, social balance of natural, authentic human society merges with biological, psychological and environmental balances to form an authentic, human ecological system. This system is an archetype of humankind, of its growth-fulfillment and perpetuation and of its evolution. Those communities which maintained it during the shift with rising population from band or extended family to tribe, flourished in numbers, and those that didn't declined in numbers. This characteristic, human, ecological system that is authentic to our species and modern subspecies can still be observed intact (in band, extended family, clan or tribe) wherever we discover upon a stone age community untouched by 'civilization' and/or overpopulation. More important, this authentic, human structure-function-dwelling is largely intact in many contemporary urban Earth-folk societies/'cultures', i.e., those not yet transmuted by the sci-technic revolution and/or communications revolutions. In tact, also, is their substantial resistance to adulterations from modern society and its automaticallized changes that are outside natural, authentic purposes and will of human nature and being.

## 17. Human Selflifeworld Collides With 'Civilization'

Authenticism reveals that the humanness of original, natural humankind–its makeup–has been unnaturalized and unauthenticized through the social and sociocultural transmutation of 'civilization', i.e., transmuted by 'civilization's' quasi 'nurture' and 'culture' wherein the whole self is separated into intellect-egoself. However, after 'civilization' collides and penetrates humankind, this quasi 'self' is found fallen out of its triecologic, social-spiritual-psychological nature, identity and autonomy; out of its unified, triecologic being of ego-self, family-self and community-self. The whole human self, having self-consciousness, observes this transmuted egoself largely apart from, rather than thoroughly intergrown with, one or both of the other portions of its triecologic self. This 'superego' is the egoself beholding and observing family-self and community-self as largely separate from, rather than naturally intergrown with it as natural, whole, human self. This unnatural, unauthentic egoself is a social orphan, suffering loss, diminishment, and disintegration of its original make up as whole self. It procedes as loner (with lonerism) consisting of biologic ego-self using inflated sight, altered other senses, diminished instinct, intuition, and inflated intellect as its mind; and, it adopts diminished, adulterated, surrogated family and community. It senses, vaguely through suppressed, subdued intuition, that these artifices are only surrogates for the missing real portions of natural, whole self. This diminished, surrogated, quasi self is the unnatural, unauthentic, quasi-human self that Freud and psychoanalysis inadequately portray. This alteration—transmutation–and disintegration of self occurs centrally within the mind, (*See* "Natural, Authentic Human Mind," Ch. V.) but transmutates lifeworld, being and spirit as well. The penetration and transmutation of natural, whole mind results in two basic disintegrations: that of the self (as described) and that of intergrown selflifeworld (as described below).

Total dysfunction of fragmented, diminished and adulterated self and selflifeworld is seen in the autistic child; total and partial dysfunction in the adult is seen in psychosis and neurosis respectively. Individual and social manifestations of alienization and artificialization of the self and selflifeworld described below include, anxiety, depression, meaninglessness, domination, submissiveness, aggression, egoism, megalomania, paranoia, schizophrenia, etc. Most of the maladies and diseases of civilized man–some possibly, notable exceptions being those of genetic, viral, bacterial, and toxic origins–originate in this unnatural disintegration and transmutation of two natural, human triecologies–the triecologic self and selflifeworld authentic to us. There is, in effect, a deterioration in, and a substantial loss or diminishment of their triecologic parts.

Recently, the emphasis in developmental psychology has been on the infant's relationship to family, especially the parents, and the struggle that ensues between the two. What is not realized is that this struggle is alien to natural, authentic human family. It arises from the child and parents (and family members) being caught between interests of their own and conflicting interests of other separated, diminished, adulterated portions of whole self; and from the presence of unnatural, unauthentic family and community–

the diminished, surrogated, subjugated family and community of 'civilization' society. A second problem joining the transmutation of self is the fragmentation and transmutation of selflifeworld. Both result from the impacts of adulterations by alien elements-phenomena, since sci-technic society has largely replaced natural, authentic human society—of selflifeworld—with disintegrative, alienized, artificialized, quasi 'socioculture'. All outright conflict within the triecologic social self, that is, among its three parts, ego-self, family-self and community-self, is from the disintegration of these two triecologies and is unnatural and alien to natural, authentic, unified, triecologic, human selflifeworld.

The dual revelation, here, then, is that the three intergrown parts of self interplay in an organismic, human, social triecosystem—of the triecologic self; a triecosystem that, in turn, interplays with a second—that of triecologic selflifeworld. Human growth, then, is merely ego centered biologically. (And, is not a new baby the center of attention?) Growing outward, bio-self acquires into itself mother/family and then community until it plays optimally in the two triecologies: of social triecologic self and of selflifeworld. The purpose of natural, human selflifeworld is its own intergrowths and creation of ego-self, family-self and community-self into a mature unity of human triecologic self with the growing triecologic selflifeworld; and the continuous perpetuation of these two triecologies, intergrowing with a third triecosystem: of body, mind and spirit, forming a tri-tri-triecologic human *being*.

As stated, self consciousness of the triecologic self is merely centered in the ego-self (from biological origins) and include, as organic parts of self-consciousness, the family and community. No one part can exist without the other two except as momentary attention or consciousness: since (a), a person removed to the wilderness lifeworld (or the alien lifeworld as described later) has lost two parts of himself and is no longer fully human; and (b), parents and offspring removed to be alone in the wilderness lifeworld or to alien, artificial lifeworld as nuclear family have lost one part of themselves, since it is prescribed by our species nature and being (of Homo sapiens) to have a lifeworld that includes community, minimally consisting of some extended family, band or clan. (The extended family sometimes functions as community, or essential part of it, for the immediate family.)

# D. 'Civilization': the Megamyth

## 18. A Fresh Wind of Freedom

'Civilization' has in our age clearly exposed itself to be a dilemma for natural, human being and all living nature.

In real, natural community, past, present and future are in general unity—in a unified field of life and being. Folklore, ceremony, tradition, custom, ritual, and other natural, authentic phenomena of nurtureculture and lifeway fulfill and perpetuate us, and tell us *who we were, are, and will be.*

We burden our lives and being with the false belief, the illusion, that the history we are taught in schools is our history. That is the big lie. It is only a story of cities and empires written by these. We humans have, by great majority (until late twentieth century), always lived in rural village, small town, or family farm wholly or substantially free of the oppression and subjugation of the dominating and wealthy breed of humanity—substantially free from 'civilization.'

Our real past as human beings—our past humanity—is hidden from us by the written word: by history; a history edited to rationalize the subjugation, oppression, slavery, and human slaughter of the people to power-wealth materialism. The 'spirit' of matter—of materialism—assaults spirit itself, assaults soul. Language, once written, is transmuted into a weapon against human spirit and reality.

The schools tell us the edited past of cities, nations and empires; it is not *our* past. It is not taught in our schools (only barely mentioned) that we come from the small town, family farm, tribal village—from rural community. Most of our grandparents are found there; and nearly all of their grandparents. Back through a thousand generations and beyond we spring from self-sufficient family and community rooted in the rich bosom of organismic Earth, Mother Nature. Even when conquered by the transmuted city-people, Homo alienus, we have usually kept alive our identity, our spirit and culture. Oppression, subjugation, even some local decimations, could not defeat our human being and spirit.

We are taught that political and military defeat is defeat; but, usually the human spirit moves around and through the hard weapons and oppressions of material power and wealth. The power-wealth breed values these alien materials; but the human spring always returns to re-grow the human spirit, for, it sprouts in unison with the spring of living nature Herself, her lifebeing force.

We need a fresh vision of freedom to displace the political visions of the democratic revolution that did rectify many of the old injustices; for, the democratic spirit has been outflanked by the exponential alienization of the industrial revolution. For 'civilized' people freedom still lies in wait to be rediscovered with the revelation that there are two distinctly different flows of life, two journeys through the valley of life: one in the flow of 'civilization' fallen from natural, pure human being, the other in the flow of natural, nurture, and selflifeworld with time and being. This revelation delivers a choice: to continue unresisting, unaware and in the infected, polluted flow of 'civilizaton' or to commence the treatment of our illness, the recovery and preservation of some of the health of natural human being. We can never be completely free and healthy human beings again, but can only resist the alien illness that progresses toward the extinction or decimation of our human being, our spirit, our soul.

## 19. The First Mistake

We, within 'civilization' lifeway, almost unavoidably and often happily, embrace our self-deception about 'civilization'; for, to be with the movement of history–the development of the material and conceptual power-wealth of 'culture'–to be in the currents of the foreseeable future deceptively seems like riding manifest destiny and a dynamic of life itself.

The megamyth and core mistake, nearly universal in the West and quite predominate in non-Western 'civilizations', are the conception and phenomenon respectively of 'civilization' as inherent givens. The misconception of progress holds for the upper and middle class until the breakdown stage of the rise and fall cycle of 'civilization'. Amidst this 'manifest destiny' of the priviledged, the enslaved and the oppressed long for their freedom and their land. Notwithstanding this illusion of 'manifest destiny', we know by accounts that at the start of the engagement of 'civilization' lifeway with natural, authentic socioculture, many Earth-folk people see through the seductive adventure-discovery power-wealth spirit-ethos of alien 'civilization'; but their voices fall on the deaf ears of the seduced. For, to be with the movement of history–the development of 'culture'–is to feel a reassurance that one's own future is unfolding as it correctly should in line with a dynamic of life itself. Written 'civilization', however, began only 5,000 years ago; natural, authentic human lives and human being (of our species) have been unfolding for 300,000 years.

Thus the original, human illusion and untruth, namely, that 'civilization' and its society, culture, and lifeway are either more good or more manifestly destined for humankind than pre'civilization' or Earth-folk a'civilization' represents the original opposition to human realty, truth and goodness; the beginning of the quasi-lifebeing-force of alienism; and it's original engagement with the real lifebeing-force of human authenticism: the beginning of the dialectical and spiritual-ontological struggle of alienism versus human athenticism. In other words, historical development and movement have been mistaken for *the* human reality, *the* human condition, or for an essential, major part of it.

# 20. Seven Revelations About 'Civilization'

Although the popular view (and, still, sometimes the academic) is that 'civilization' and human culture (and socioculture) are synonymous; in actuality, as social science has come to hold, 'civilization' is but one 'species' of 'culture' and socioculture. Authenticism reveals this new, later variety of quasi 'culture' to be a departure from the natural, human culture of our 50,000 year old subspecies; and reveals several truths about Homo sapiens sapiens *vis-a-vis* 'civilization'. Seven will be enumerated here in Parts 22 through 28.

# 21. Written History's Misrepresentation of Humankind

One, 'civilization' is not what it portrays itself to be–the general condition of our human species. For, after all, our species lived 300,000 years, and our subspecies lived about 45,000 years before even the scattered appearances of the city and state emerged on human terrain. By one educated estimate, from 5,000 B.P. (Before Present) to 500 B.P., the percentage of humans within 'civilization'– or 'civilization' lifeway–ranged with fluctuations from about .1 percent to about two percent. By 500 B.P., only perhaps two to four percent of humankind including the West (*See* below) were living the 'civilization' lifemode, even though perhaps another ten to twenty percent were being subjugated by their urban, 'civilizationized' oppressors. It should be stressed here that to be subjugated by a state or empire falls short of being conquered, when subjugated people continue their Earth-folk lifeway whilst burdened by a tax on their production or labor. By 100 B.P. only about five to ten percent of humanity were living the 'civilization' lifeway; by 10 B.P. about forty to forty-five percent. When 'civilization' man became or becomes a majority depends by one criteria on the point at which the large town becomes a small city, i.e., on whether the line is drawn at 5,000 or at 10,000 population or somewhere between. By another consideration, it depends upon the point in time when a Earth-folk community ceases its customs, values and purposes, and is substantially overwhelmed by the propagandized consumerism, materialism and power-wealth spirit-ethos.

## 22. 'Civilization' as Humanity's Misbegotten Socioculture

Two, it follows from the above first truth that anything–in this case 'civilization'—that has been generally or thoroughly absent from a species throughout it's duration cannot be an essential or defining characteristic, i.e., cannot be within the nature or general, human condition of the species, cannot be within it's nature, being, identity and reality. By this measure, 'civilization', a recently emerging, artificial 'species' of 'culture', is, thereby, illegitimate to our species. Our humanness is confined to behavior and phenemena characteristic and authentic to our species and subspecies; it excludes behavior and activities peculiar to 'civilization' socioculture. (*See* natural, nurtureculture above and below in this chapter.)

## 23. Who's In and Who's Out

The third revelation is that education's view of the past falsely groups all people within the *geographical boundaries* of a sociocultural system, a particular 'civilization', as being, thereby, within it's 'civilization' lifeway or socioculture. This notion even contradicts the traditional definition of 'civilization' that entails such defining characteristics as literacy, heterogynousness, social complexity and urbanism, all of which start at settlements that have reached roughly 5,000 or 10,000 people. Urbanism is of singular importance because it generally entails all other defining characteristics. Beyond the claimed geographical territory of a particular 'civilization', it, nevertheless, still includes sanctuaries–beyond the cities' limits–rural people(s) that are in pre-20th century times (with very few exceptions) still living Earth-folk socioculture lifeway. They are, in effect, generally autonomous families, communities, clans and tribes; whether they are settled in family farms, small towns or communal villages, or are nomadic; whether they are hunter-gatherer, pastorial, or agricultural. Such peoples, therefore, have been and are, contrary to popular perception, outside 'civilization' lifeway. Rural people and subjugated, ethnic cultures have largely dwelled throughout historic times–and before the Industrial Revolution–in rural, Earth-folk socioculture. Presently, many are still largely free of the Industrial and Communications Revolutions and still dwell in a'civilization' Earth-folk socioculture. It is noteworthy that rural people of Earth-folk socioculture have commonly contrasted themselves, at times, neutrally, to city people with the term "folk" as in expressions like "Out here, we're just 'folk'.", and, at times, favorably with city people by their reference to them as 'educated fools' or 'city slickers'. These Earth-folk of a'civilization' socioculture decline in numbers as some are 'civilizationized' when the communications, transportation and industrial revolutions reach into the Earth-folk socioculture. A convenient time or rough demarcation for these penetrations of Earth-folk socioculture by 'civilization' is the first half of the 20th century. Rural folk, thereby, are now being adulterated by the process of modern 'civilization'; in the First World to the extent of being absorbed; in the Second World Earth-folk socioculture is in full conflict with 'civilization'; and in the Third World or pre-industrial world, Earth-folk socioculture predominates. To say that the process of so-called *progress* 'civilizes' human society is false; for, it has always been generally civil. It is, instead, being 'civilization'ized.

## 24. The Dichotomy of Human Society

A fourth revelation about 'civilization' and humanity is that this division of our 50,000 year old subspecies into two distinct, opposing sociocultures is a crucial, requisite revelation, a primary human dichotomy that, unlike most dichotomies, is indissolvable; and, it is also indispensable to our understanding if humankind is to recover and unify some of the lost portions of our natural, authentic human nature, being, identity, spirit, reality and selflifeworld. The matter of further sub-division of this primary division is of secondary importance, some examples being, whether Earth-folk socioculture should be divided into pre-Neolithic and post-Neolithic, into hunter-gatherer and agricultural society, into nomadic, pastorial, and settled society, into Paleolithic, Neolithic and post-Neolithic society or into Paleolithic, Neolithic and a'civilization' Earth-folk societies. Regardless of any such elected subdivisions, the basic requisite dichotomy is and must be understood as: 'civilization' socioculture versus non-'civilization' socioculture (the preferred terms, herein, being a'civilization' or Earth-folk socioculture).

In this volume, the preferred further subdivision of humankind is into three parts—pre-'civilization' Earth-folk socioculture, a'civilization' Earth-folk socioculture, and 'civilization' socioculture. However, the crucial, basic dichotomy remains and must be understood as 'civilization' versus a'civilization, Earth-folk socioculture; this is more easily realized once one abandons city, state, and 'civilization' centered perspectives—and the quasi-human values they contain—when reflecting on matters relating to the nature of human being, reality and their selflifeworld.

## 25. The Endurance and Triumph of Unwritten History—Folklore

The fifth revelation is that there is another historical account countering the one written by conquerers; and, afterall, what was conquered or fragmented was freedom—organic family-community autonomies. 'Civilization' people must abandon the conceited, condescending and false belief that the written history of his own 'civilization' (socioculture) and that those of other 'civilizations' are the only histories that are or have been in use by humanity. In fact, written history has been both a misrepresentation and misappropriation of the human past. It has been a history of a small minority, of the 'civilization' sociocultures that have in their very nature, key controlling and directing powers such as that of the written word, of formal institutional education, as well as, powers of the military, commerce, and industries. It is only contemporarily that this written history of the small power-wealth, exploitive minority–a history of the conquests and retreats of power-wealth itself and of the power-wealth spirit-ethos–now makes its intended, final bid, after 5,000 years of growing amidst the majorities oral, Earth-folk history, folklore and mythology of Earth-folk a'civilization' socioculture, to become the majority history.

This bid is now being blocked–beyond the traditional checks of Earth-folk socioculture's resistant culture, e.g., folklore–ironically by some of the cityite's own sciences, mainly anthropology, and by some of its own industries, like revisionist films, videos, and television. Indeed, then, let *this* word (oral and written) go forth to both peoples of humankind: (a) to 'civilization' sociocultures where natural, authentic human nature, being, identity and reality are progressively blended with adulterated elements-phenomena of 'civilization', and (b) to Earth-folk sociocultures where natural, authentic humankind is only incidentally penetrated, where perimeters are holding against the threatening influx of adulteration–that the true, real and authentic history of our human species has always reigned in Earth-folk communities over the lying written word of the powerful and wealthy; that it has reigned in the greatly predominant Earth-folk socioculture and lifeway of *us*, of modern humankind; that it has reigned via the oral story teller and epic singer, and via the rest of folklore: customs, beliefs, ritual, dance, ceremony, Earth-folk music and other non-literary and literary manifestations. And, let another message go forth from human authenticism to all humankind, that both our true and authentic history, and the true and authentic human nature, being, identity, reality and selflifeworld of these *are* holding forth with renewed hope; and that the revelation of human authenticism, being a complete system of revelatory philosophy, foretells of a plausible continued reign of our 50,000 year old, authentic, human society over 'civilization' society: foretells of a plausible triumph of natural, authentic, human spirit over the faltering, incredulous, unsustainable, de-evolutional, 5,000 year-old power-wealth spirit-ethos.

## 26. Written History as Elitist Myth and Propaganda

The sixth revelation is of the falsehood and illusion that written history is inherently vested in. Written history is 'civilization's' big falsehood or myth because it is the given account of a power-wealth minority class, growing from a miniscule percentage at its emergence to approach a majority class only contemporaneously; an account of that growing minority class, given by itself of its diverse sociocultural surges and subsidings. Can an accurate investigation and truthful account of any event be made by involved parties with vested interests? All written histories are laughable exercises in cultural arrogance and embellishment to any unsullied, independent citizen-philospher of humankind. Only of late has the history–past–of the majority of humankind *begun* to be put into writing, and very largely, erroneously, because most of such writers embrace generally the power-wealth spirit-ethos: they propagate the central lie of both written history and 'civilization'– that 'civilization' society is superior to Earth-folk society.

## 27. Historic Myth and Self-illusion Become Species Delusion

Seven, 'civilization's' megaillusion is intertwined, then, with it's megamyth, it's written history, it's self-illusion. The composite errors and false values of 'civilization' that swirl around the core of this megamyth/illusion–the power-wealth spirit-ethos–have been partially mentioned in the above revelations about our species, modern man, and 'civilization'. Other, remaining, composite flaws are revealed throughout this volume. The basic, redeeming consolation is that this alien power-wealth ethos and system, from their emergence onward, are countered and opposed by the natural growth/perpetuation sufficiency of natural, authentic, human being, nature and spirit: by the lifebeing-force of human authenticism.

# E. Science Out on a Limb

## 28. An Age Surprise for Humanity

Historians and nearly all of academia are now, in our time, suddenly caught out on a limb, out on the high branch of 'civilization' socioculture, thinking and having thought from its emergence that this branch is the trunk of our tree of humanity, or at least the central branch up from the trunk. They and we are on a weakening branch, caught and exposed by the increased light thrown upon it and the branch below–primordial, authentic humanity–by the human and animal sciences. Amidst this new light all bets are off—all theories, suppositions, givens, ideas and beliefs–all are up for disposal or radical revision. This new awareness is on the threshold of a greater final one: the human revelation of human authenticism, which reveals this branch of humanity as decayous and diseased by its penetration and pathogenesis from artificial, alien elements-phenomena.

Perhaps the central misconception blocking us 'civilized' people from understanding ourselves has been the *traditional* presumption of recent emergence of the human group (species or subspecies) to which we identify. Our species is now dated back to about 300,000 years; our subspecies to about 50,000 years. Heretofore, however, lacking recent archealogical findings, it has been natural to select a satisfying and self-flattering probability or estimation: that 'civilization' man is the developed, mature version of humankind; with previous ones and Earth-folk socioculture being embryonic and childish versions, respectively. This perspective is forced upon the historian and the educator from their preoccupation, from vested interest (personal, professional and sociocultural) with the supposedly greater importance of 'civilization' socioculture over pre'civilization' and Earth-folk socioculture in the pursuit of understanding humankind, our nature and being (collective and personal).

The new awareness that 'civilization' is a recent, minor part of our species life and history, along with the revelation that it is decayous and alien, generally devastates 'civilization' ideologies and presumptions. Recent knowledge utterly destroys many foundational or core ideas of 'civilization', regardless of a dangerous need felt by many to cling to comfortable falsehood and collective self-deception and self-dilusion.

## 29. The Myth of 'Barbarians' Vs 'Civilization'

One such foundational idea of 'civilization' destroyed herein is that a civilization' or Earth-folk sociocultures remaining outside of or free of the conquest, servitude, subjugation or enslavement of city-states, states, nations or empires, are barbarian sociocultures. This is hardly more than ethnocentric egotism, and merely a key propaganda line of conquering sociocultures, which is implanted into their written histories. This preposterous idea becomes dogma by serving to justify and even sanction past and future conquests. In the endless socioculture marathon of war, the competing enemy is always the quilty barbarian: for, history is one of the spoils of war and records both the incursionary challenger and the previously routed as barbarians; when, actually, the game is power-wealth and the call, "Who's turn ?". By the logic of this game a people must conquer another people to escape the condition and label of barbarianism, become 'civilized' or re'civilized' and certify itself as such. In truth, the rural communities and tribes of humankind bequething nurtureculture sufficiency-growth-fulfillment-perpetuation from generation to generation have freedom and welfare so long as they, like non-human animals, avoid the physical, mental or spiritual capture of conquering socioculture–'civilization'.

This destroyed, foundational idea emerged and was sustained, also, by the mere conflict of two social groups and could be termed the *them vs. us* phenomenon, which, when attained to battle, renders 'them' as barbarous in their style or method of battle.

## 30. Equestrian Warriors as Aberrational

A second, foundational idea of 'civilization' destroyed by the revelation of human authenticism is that the human of pre'civilization' and non'civilization' was/is a savage–even when granted to be a 'noble savage'. This idea has been thoroughly discredited by anthropology and the other human sciences, but is still clutched by the collective ego of 'civilization' consciousness and world view thought, especially upon occasion for self-defense. Two primary notions falsely support this myth of a primordial savage: (a) that prehistoric man was/is a solitary, a loner like the mountain lion or bear, with an aggressive ego-self: in truth, he was/is of extended, family communities of one or several extended families or clans. And (b) that he roamed in bands ever hostile or at war with other primitives: in truth, this phenomenon occured only incidentally and outside the general pre'civilization' behavior patterns. The truth is that both 'civilization' man and a'civilization' man is provoked to savagery, generally, by the same kinds of events.

Unfortunately, the 'savage myth' gains some support by the well established fact that nomadic herdsmen of pre and a'civilization' sociocultures conducted sporadic and, in cases, regular raids of pillage and conquest on peaceful, settled agrarian peoples and peaceful nomadic hunter-gatherers. Several pages could be devoted here to the alienized and distinct a'civilization' socioculture of equestrian nomadic herdsmen. It will suffice here, to reveal three characteristics of equestrian nomadic husbandry which render it alien to and outside of natural, authentic, human socioculture. Equestrian, nomadic husbandry is alien socioculture, first, because its heyday existed only about 3,000 years; second, in that heyday it zenithed at about 15-20 percent of human population; and, third, less than one or two percent of modern human beings, i.e., of humans from about 50,000 B.P. to the present, have mounted a horse for the purpose of nomadic herding. Nomadic husbandry itself becomes generally troublesome only when of the equestrian form. Nomadic husbandry on foot is exceptional and distinct–an adaptational lifeway within natural, human, adaptational experiment, destined by the alien phenomena of progressive over population, both to remain outside the general human nature and reality of our species' human condition, and to merge with another alien phenomenon, the mounting of the horse by humans.

## 31. The Precursor, Alien Society

This equestrian pastoralism, indeed, is probably the greatest example we have of precursor, nongeneral, alien, quasi-human society. In prevalence, it was, at the zenith of its c.3,000 year heyday, perhaps in third place as it vied with hunter-gatherer, settled agrarian, and (in fourth place) emergent 'civilization' socioculture. Some nomadic herdsmen–the Germanic–for centuries challenged, with substantial success, the Roman empire, forcing the defense of both its socioculture (cities) and subjugated or hegemonized, rural, Earth-folk sociocultures. Here, an interesting question arises to us, as it must have to threatened Earth-folk peoples during this period: namely, which of the two threatening, alien lifeways–herdsmanship and 'civilization's urbanization–posed the greater threat to the two natural, authentic sociocultures, i.e., hunter-gatherer and settled agrarian. More precisely, which was the greater of the two evils? Historically, 'civilization' socioculture proved its greater strength over its competitor, equestrian nomadism in the competition for conquest. Though on the personal, temporal, level, records indicate it was a close call: many rural people(s) forced to choose between oppressive, decadent, Roman urbanism and the less complex, expansionist, threatening, tribal herdsmanship, chose the latter as the lesser evil.

## 32. The Horse and Dog as Suspects for Alienism

Because horseback riding emerged approximately concurrent with 'civilization' (c.5,000 B.P.)–or as later evidence suggests, 7,000 B.P. prior to 'civilization'–as distinct from domestication for food (c.9,000 B.P.); it is, therefore, another suspect as the emergent point of alienism. All the more so because the mounting of another species for warfare and human transportation is both a more substantial act of enslavement of the horse than mere confinement to pasture; and, as well, a significant renouncement of natural, human biological ambulation—walking. Various species are known to freeload on the bodies of a different species, but in all such cases, the 'host' remains in charge of its natural affairs. Not so the horse with equestrianism; wherein the horse is subjugated into the human battlefield, ceremonial and transportational purposes, which are clearly alien to the horse's nature, being and species purpose. Horseback riding, except as recreation, is clearly an alien element-phenomenon within humanity. The only alternative conclusion would seem to be the insupportable proposition that horse nature, being, and reality has intergrown with those of humankind, a proposition that is notibly represented in the Greek myth of the Centaur.

Indeed, the attempt by diverse 'civilization' sociocultures to give the horse a natural and authentic place within the human condition and its socioculture is best illustrated by this mythical Centaur of the ancient Greek religion of a 'civilization' that proposed and sustained in the consciousness the belief that the Centaur had once lived as human/horse beings, and had conveyed to human heros pivital teachings. This attempt was doomed to failure because, aside from its enslavement into domestication breeding, the horse is born free into its own natural, species being and would live so except for the quick interference after birth of horse domestication. 'Civilization' horse, like 'civilization' man, is imprisoned within alien socioculture. 'Civilization's breeding of horse and man does not make for the natural growth of their respective beings; but, instead, it makes for their diminishment and decay, through adulteration and violation, of their social and mental nature, and of their being and spirit.

In this volume, the nature of the relationship between dog and man will be limited. The relationship between the two is clearly alien once it is within 'civilization'. However, if the dog became the general hunting partner of man—in a preponderant percentage of human communities and for over half of our subspecies human existence (for over 20,000 years)—then canine and human nature, being, and reality did become significantly species intergrown and interfunctional, thereby, making the dog within a 'civilization' Earth-folk lifeway a natural, authentic element of our human subspecies and its selflifeworld. If, on the other hand, this relationship did not achieve the species intergrown magnitude, the phenomenon itself remains a natural experiment to incorporate a natural element-phenomenon, the dog, into our human species nature, being, reality, and selflifeworld. That is to say, it remains, in the latter case, a significant incidental, rather than a general, characteristic element-phenomenon of our species.

If one concludes, with or without future evidence, that general horseback riding emerged as the first alien element-phenomenon penetration, before the more general and comprehensive alien phenomenon of the city emerged; then there remains room within the philosophy of human authenticism for such refinements of proposition without damaging essentials of the general revelation and philosophy of human authenticism.

It should be stressed, here, that the issues of *when* alienism first penetrated human authenticism, and *what* was the first, alien element of penetration, are not main or crucial issues to the mega revelation of human authenticism and its philosophy. These two issues do not effect the revelation that the condition of human authenticism exists and once was the sole and original authentic human condition until it was penetrated by alienism, the *general*, substantial manifestation of which is 'civilization'.

## F. Other Possible Origins of Alienism

Hopeful, adventurous, and ready for rediscovery: this describes our outlook after receiving the revelation that 'civilization' is alien and artificial to the necessarily natural nature of human being; for, it marks an upward turning, a point of healing for 'civilization'. And, it presents quite enough exciting challenges encompassed in the working out of the details within the consequent movement into human reauthenticization, into renurturalization.

One such detail is the question of what essential and vital divisions, other than the one affirmed herein, could be, hypothetically, drawn within the c.50,000 year span of Homo sapiens sapiens and the c.300,000 year span of Home sapiens *vis-a-vis* our purpose herein of defining the natural, authentic, nature, being, identity, reality and spirit that are to be realized as belonging rightfully to/with contemporary humankind. Should our central focus include our 300,000 year old species, our 50,000 year old subspecies, and/or our period from the start of settled agriculture (c.10,000 B.P.) to the present? For 'civilization's main schools of thought it is a mute question, with their answer, "none of the above": these thinkers base their thought, activities, and perceptions on the history of 'civilization' socioculture, and all those who have fallen into 'civilization's egoism of intellect-egoself.

However, given our revelation, a few other possibilities beg some consideration. First, with 'civilization' revealed as alien, may not the Neolithic stage be included into this alienism, thereby marking agriculture as the initial, alien penetration of natural socioculture–nurtureculture; and, second, may not the entire 50,000 year old subspecies, Homo sapien sapiens be, in fact, an alienized subspecies commencing its progressive movement and divergence out of a wholy, natural species–Neanderthal man (about 300,000 years old) and its natural socioculture. The second question spins out from its false presumptive ground a series of less plausible and even nonsensical possibilities: e.g., that our entire human species (c.350,000 years old) is alien; that our genus is alien; that our human family–*human being* itself–is alien to living nature and animal socioculture; that primates are alien; that mammals or all land dwelling animals are alien because they appeared relatively late from sea life; and, lastly, that life itself is an alien phenomenon divergent from true authentic living nature: inanimate matter.

However, if living nature and participating inanimate nature are both inclusive to living nature; and if living nature is more relevant to our nature and being, then all of the above possibilities can be addressed by addressing the first three. Any theory of a dialectical conflict between living nature and unliving nature entails a penetration of living nature by an object or phenomenon of alien nature–by an artificial, alien element/phenomenon. Bipedalism, the growth of the human brain and most of the resulting changes in lifeway are clearly within biological evolution–within the first, dynamic of living nature: natural selection and mutation. Some others, such as language and tool use, authenticism gives no plausibility as suspects, as agents of artificial, alien penetration; and several have already been addressed in Part B of "THE REVELATION:".

As for the possibility that Neanderthal man was infected for over 300,000 years with alien, pathogenic phenomenon, this extremely unlikely prospect could take at least two forms. The pathogenesis could have effected slight harm and been held at steady state. And, then too, it could have been contained in a rise and decline cycle, similar to one such of the cycles in 'civilization' socioculture, never becoming decisivly damaging, or only upon the addition of 'civilization'. As for Homo sapiens sapiens of the last 50,000 years, these two possible scenarios for penetrated Neanderthal would apply, in the same manner. Nonetheless, both traditional 'civilization' man and recent, sci-technic man should be so fortunate, infected as they are with cyclical progression away from human being and nature, a progression into decay and decimation.

There is, in fact, some attraction and possibly some merit in the idea of drawing a divisionary line between the hunter-gatherer lifeway of our 300,000 year old species, and the agricultural lifeway which emerged only c.10,000 years ago as the first basic alternative lifeway. However, in doing this, a second divisionary line cannot be avoided since a second, basic alternative emerged c.5,000 years ago and is not just a radical lifeway change from agriculture, but a change more radical that the change from hunter-gatherer.

This is especially true about drastically artificialized, surrogated and adulterated contemporary 'civilization' lifeways. Joining this first, there is a second, main obstacle to a single division of our species into pre and post agriculture, having agriculture as the first alien penetrant of our species' natural socioculture–nurtureculture: it is the length of time of this human stage and the large percentage of diverse peoples that it encircled. About 90 percent of present humankind–by its ethnic, tribal, linguistic, and sociocultural groupings–have emerged from about 3 to 6 thousand years within a general, rural, agricultural lifeway. This duration and this extent of our species recent experience within the farming/animal domestication lifeway grants this lifeway a significant (though small), contributing role in the make up of humankind's present collective nature, being, identity, reality and spirit. A role, still though, far short of that of the hunter-gatherer lifeway. Contrarily, 'civilization' lifeway only contemporarily encircles about 60 percent of humankind; and being generaly alienized and artificialized has departed from and is in conflict with earlier sociocultures of living nature's primates and humans, and cannot ever be incorporated into living (human) nature, being and reality, which grows, fulfills and perpetuates; but can only be incorporated into nature and being as the agent of decay. 'Civilization' cannot *live* for it is not organic, but mechanic. Only its sociobiologic tissue *grows* as life infected with alien, artificial pollutants in the form of pathogenesis.

The case for agriculture and husbandry as the first, minor but significant penetrant is seen, herein, as better than the case for it as the general penetrant of natural, human socioculture; it is the second best proposition to the proposed one in this volume, i.e., that the later form of city-'civilization' is the violator. Notwithstanding this proposition, it is fairly clear that the traditional family farm and village community–pre-twentieth century automobile and radio mass production–show significant but no substantial destruction of triecologic self, family, community and selflifeworld. Hence, and in summary, there is room within the philosophy of human

authenticism for other, earlier significant and/or incidental penetration prior to general penetration of natural socioculture by
'civilization' of the Sumerian type.

The basic fact remains that any breakthrough revelation transforming 'civilization' man's perception and conception of himself
and human socioculture will illuminate two fundamental human
sociocultures with great, conspicious differences between them,
and will reveal them as dichotymous, antithetical and engaged in
dialectical struggle. That is to say, our redemption and salvation
must be through a core revelation about both 'civilization'
socioculture and a'civilization', Earth-folk socioculture. It is of only
incidental importance if the initial divergence from living nature
did indeed occur at an earlier time than emergent 'cilization'; for,
the significant, manifestation of divergent alienism is 'civilization',
the oppositional, cyclical, progressive socioculture that disintegrates
the natural, nurturecultural sociocultures, and lifeworld that are
original and authentic to our human species and subspecies.

## G. Selflifeworld as Subject-Object in Unity

### Short Forward

There is only one primary difference between subject and object. Subjects behold or experience objects. Objects cannot behold
or experience; they have no life; subjects have life. Some objects
living nature pulls into a participation in the lifeworld of her manifold subjects. Living nature utilized them in her evolved web of
life—her Earthlifeworld. These are subject-objects or life-objects.
All other objects are left out of living reality—mere objects only.
When we ponder or utilize them, we adulterate, diminish and
alienize our living being and its selflifeworld.

The two revelations, of triecologic self and of selflifeworld, can be
stated in philosophy's perspective of subject and object. Succintly
put, the natural, authentic, human self of our species is not ego self as
subject, but rather triecologic self as *subject*. What has been thought

to be it's most immediate *objects*, family and community, are not objects to the socio-biological species including modern man. These two are, instead, two parts of the self intergrown with the third part, bio-self, all together constituting triecologic subject; whereas bio-self is self only in biologic being and nature. Bio-ego, as mind-body organism only, does behold these 'objects', but observes them not as objects, but as parts–subject-objects–of itself as subject-object, just as it observes its body parts as parts of its biological self. That is to say, human bio-ego-self, as the biological part of triecologic, social self beholds family and community and its entities as subjective parts of itself. Family and community entities have long since evolved, along with social, animal emergence, into essential parts of social animal self. They are objects evolved into subject parts, that, along with an ecosystem's field of beings, make ego into whole human self, just as, in parallel, humans are biological animals evolved into sociobiologic animals. They are subjects-objects parallelling humans-animals.

*   *   *

Arriving here at subject-object, we must abandon the traditional view of subject-object dichotomy: this is an untrue and illusionary view of human reality and being. It is only the biological self–along with the intellect egoself as explained in Ch. IV, EGOSELF, ORIGINS AND DEVELOPMENTS,—that beholds objects; the socio-biological-spriritual-ontological, *human* self beholds primarily subject-objects–it observes mother/bosom just as the subject-object it is, a part of its socio-biological-ontological self as subject-object. And, it observes itself as subject-object. That is, in whole human self (socio-biological-ontological self), self is subject-object. Bio-ego 'self' (subject) is biological only—pre-social being; it is not self-conscious and can only behold the objects outside itself. Self-consciousness orginates self as subject-object and is socio-biological-spiritual-ontological selfhood including *human* self.

More fully, the natural, authentic, human self observes bio-self as subject-object, family (mother, siblings, et al) as subjects/ objects and community members as subjects/objects; it observes these as self-conscious observation of whole, triecologic self consisting of bio-self, family-self and community-self. To this natural, bio-social self, objects are still incidental rather than having become general oppositional phenomena, as becomes the case with 'civilization' man. Since life is a part of the living self, and life and self occur in their natural world, lifeworld objects are also parts of the human self, intergrown in selflifeworld. Is man, after all, still man without his tools, utensils, clothes, plants, soil, animals, et al? No, they are essential constituents of his species identity, nature, being and reality. While the human self observes it's lifeworld objects as subject-objects—as subject-object-objects—or as subjective objects (subjectified objects), only that part of the world which does not touch essential, human selflifeworld is pure object. Objects are world objects observed but remaining outside subject-object—outside selflifeworld. *Imagined* objects, world objects and subject-objects become real only through sensual contacts that incorporate them into the human beings' selflifeworld, nature and being (e.g. seeing, touching, hearing, smelling and tasting). However, sensual contacts that are motivated by curiosity or experimentation alone remain unreal: until they are correctly and fittingly incorporated into the living nature of a living being and its selflifeworld. Otherwise, they remain alien objects that in alien, artificial 'civilization' often are of continuing curiosity and experiment. Thus, selflifeworld is subject-subject-object/subject-object/ subject-object-object.

The revelation here is that self/world and subject-object are continuums, not dichotomies, because they are, in reality, selflifeworld. The lines we have drawn on these continuums that separate and oppose the spheres are drawn by 'civilization' and unnatural, 'civilization' man. In natural, authentic human self, lifeworld and nurtureculture there is no pure subject, no pure object. In unnatural human lifeworld (i.e. 'civilization'), unnatural man has created pure object i.e. objects that have no significance to natural authentic human self and lifeworld. Pure object is the alien, that which is rejected as not of and belonging with natural self and lifeworld–rejected by natural, authentic man; but accepted and even created by unnatural, unauthentic blended, synthetic man, having become partially synthetic and alien, trapped in progressive alien life and world; his debilitated will becomes a contributing force in the creation of unreal self, life and world and distructive of real, authentic human selflifeworld. Object, of course, does not exist in self-conscious selflifeworld without subject: they are intergrown as natural selflifeworld; object, however is separated in unnatural, alien ego self, life and world. This creation of object was simultaneous with the creation of pure subject, the egoself—the ego seperated from family, community and lifeworld. Freud partially realized this egoself's existence and inadequately understood it, calling it super-ego. It is actually a mutant, alienized, quasi self. (*See* Ch. V, NATURAL, AUTHENTIC, HUMAN MIND, and Ch. VIII, HUMAN AUTHENTICISM Vs. ALIENISM.)

For an entity to be a human being, it must be living its distinctive lifeworld; otherwise, it is either a corpse or a living, quasi-human, mind-body organism confined through artificial devices away from its lifeworld by other quasi-human beings. Removed from the lifeworld of its species being, the entity's being is diminished, adulterated and transmuted by the alien conditions of its confinement into partially, nonhuman being. It is alienized into alien/human being. Therefore, there is no such thing as a natural, living being *as a subject* removed from the objects that help create and sustain the nature of its being and nature. To be a living being is to be subject-object–more specifically, subject-subject-object/subject-object/subject-object-object.

Subjects are created within 'civilization' by the separation of human beings from their essential human objects–subject-objects–replacing them with alien, quasi objects, objects that do not belong with humans and cannot intergrow and interfunction with the growth-fulfillment of human being.

The newborn baby is essential ego suspended, momentarily lost in pure ego, desperately seeking selfhood and lifeworld in the union with mother at point and time of contact—upon union with bosom and mother. At that union ego becomes self; it becomes subject-object intergrown to subject-object (mother), intergrown as newborn whole self. This newborn social self, limited to mother, is essential, newborn subject-object, becoming triecologic self as the child relates to father, siblings, peers, et al. Until age three or four the family serves a duel role as family self and rudimentary community self while the early childhood self is growing and developing it's community self which is substantially formed by about age five.

The newborn having grown/evolved from an embryonic being, emerges from embroyonic, fetal realm and desperately seeks and forthwith finds (barring interference) it's way into it's lifeworld with mother/bosom union, finding and creating, at once, self and lifeworld. For this human infant, from the beginning until development of family self and community self, the human self (subject) *is* self/mother (subject/subject-object). For our species as with other sociobiologic species, e.g., previous human species, primate species, and social mammals, there is no distinct, separate, independent ego self. Such an ego self is an illusion arising from the awareness of the biological, ego portion of us which is only a part of an intergrown biological social self–a unified, whole human self. It is within an unnatural, alien, human socioculture–'civilization'— that natural, whole, human self is fragmented into quasi, alienized egoself, alienized life and alienized world. At birth the midwife holds a biologic ego struggling toward and in search of self. More fully: at birth there is momentarily biological ego (essential object) struggling and searching toward human biologic-social self (essential subject-object) which is for human species and all social species the true, natural and authentic self. In human and primate species a third self has evolved from rudimentary beginnings within family, and family self into community self and community creating triecologic self.

# CHAPTER IV

# Egoself: Origins, Developments, and Arisings

## A. Organismicities: Whole Self and Human Being

CONTRARY TO THE popular, Western view of human being as an independent, atomistic subject, human being is an organismicity within the organismicities of selflifeworld and Earthlifeworld. Biologic, human being–mind-body being and egoself being–is not coextensive nor the same as human being, which exists as the organismicity of bio-self, family-self and community-self intergrown with two other triecologic organismicities, i.e., human selflifeworld and Earthlifeworld. There is, in reality, no individual human being, precisely and fully speaking, the individual is only an intergrown physical and mental part of human being. Individual being cannot exist (live) alone, except mementarily; therefore, it cannot be lone being, except momentarily. There is no independent, autonomous human being. The individual human being is a partial (part of) human being, strictly speaking. Human mind-body is merely the 'organ' of human consciousness and self-consciousness–an entity of conciousness that when naturally conscious (free from transmutation by artificial socioculture) is aware of itself as only that, and also aware of human being in which it participates.

Reality and truth, united as one, yield a paradox: that any

bio-socio-spiritual-ontological being including human being it-self, in order to be with its living nature and reality, must be intergrown and interfunctional with other entities and their natural social units, physical units and their ecosystems. That is to say, human being is intergrown, organismic entities: it is the organismicity of selflifeworld; intergrown with Earthlifeworld, (termed *biosphere* and *ecosphere* by science of ecology).

## B. Egoself: Heidegger, Fromm, Psychology and Individualism

At the outset–and leaving the experts aside for a moment–the evidence that their is a common aberrational form of self shouts out at us in four conspicious phenomena. First, we know that there are people amongst us that have larger than normal egos, that cannot function without defending these large egos that fit their lifestyle (lifeway), and are indispensable to them: the deflation of the oversized-ego brings on a crisis or a breakdown. Second, we know that intellectuals amongst us acknowledge the term, *intellectual,* and the swollen, highly developed intellect it conotes. Intellectuals have undergone an inflation of that quasi portion of natural, real intelligence called intellect, at the cost of intuition, instinct, emotion, revelation, and natural sensings. These intellectuals are as distinct from the general, human population with their natural, balanced intelligence as egotists are distinct from the general population's balanced self and personality. The issue is whether they are a superior breed or representation, or instead an alienated, deficient, and *alien* one.

A problem concerning the ego has been perceived in recent times by several eminent thinkers. Heidegger came to believe the West needed to be freed from the grip of the ego by a breakthrough and change in consciousness. Erich Fromm held in his later years that egotism had become in the West a massive barrier to human fulfillment. Recently, one school of pop psychology, lead by John Bradshaw and others, recognizes this ego problem and targets the ego as the "false self", proposing to cope with it by a therapy toward supposedly reclaiming the child within us and this child's resilience and flexibility, thereby achieving a regeneration or fresh start. The flaw here is that the goal of pop psychology is inherently deficient and false: achieving a successful adult, personal or individual identity. It is partly because individualism has been pushed beyond its beneficial role into diminishing returns and human destructivity, through the eclipse of whole self by egoself, that such coping mechanisms are needed and sought. (*See* "Individualism," Part 32 of Ch. X.)

We have in the West mistakenly equated, in general, the ego with the self, and quite consistantly. In Michael F. Zimmerman's *The Eclipse of the Self,* (Athens, OH: Ohio University Press, 1981), the author equates the self with the ego both in the title and in the book; as does Heidegger whom the book is about. He quotes Heidegger as saying in effect that we all must free ourselves from the self or the ego so that we may experience our being. Maslow toward the end of his life concluded that "people need to transcend the ego, the self, the identity, to go beyond self-actualization." Transpersonal psychology, generally, has abandoned the self as single, independent entity.

Contrary to the above ideas about the ego and the self, authenticism holds that when the natural, authentic, whole self is rediscovered, reunited and reinstalled in its rediscovered, reunited structure-function-dwelling of authentic selflifeworld, our being, along with its nature, spirit and reality are therein provided. The fallen self, the ego (the ego-intellect), a fragment of whole, human self, can provide each and all of us only a fragment of the human being authentically ours (Homo sapiens sapiens) for about 50,000 years and much longer as a species.

In this chapter, then, the problem of the ego is resolved by revealing its roots to be in the doctrine of individualism, and in other flawed conceptions addressed in this volume. The problem stems largely from the false idea that the ego and self (or subject) are generally synonomous; this despite Freud's revelation, though incomplete, that ego and self are not the same. Since human ego (biologic ego), mind and being (and the nature and reality of these) necessarily include family and community in order to exist (to be as their natures constitute them); and since human beings are not just biologic but also social and spiritual beings; it follows from either one and both of these premises that human being cannot be defined as the existential state or phenomenon of a singular entity such as egoself or subject. Instead, it is merely the senses and the intellect's awareness of them that perceive the egosubject that they serve as the dwelling place or structure-function-dwelling of human being; while the balanced, integrated, (organismic) whole mind, before its penetration and transmutation—when in its natural state–knows better.

More precisely, the manifest intelligence of this natural structure-function of mind knows that mind-body being is only a portion of human being that exists or dwells as its natural stucture-function-dwelling constituted as triecologic self (bio-self, family-self and community-self) intergrown and interfunctional with natural life and natural world–selflifeworld, and intergrown also as mind-body-spirit. This is tri-tri-triecologic human being.

Any psychology renouncing a floundering and dysfunctional adult egoself of low egoself-esteem and embracing the recollected and restrained child within, is merely utilizing a natural coping phenomenon (mechanism) usually achieved without a therapist and therapeutic psychology: it is merely good, natural healing or good therapy, respectively. Moreover, this psychology rests upon a false presumption about the child person; the result is that in time, it and its patients run head-on into a contradiction and a wall: into the reality that the resilience of childhood is inseparable from its essential preconditions, one of which is a dependence upon guidance from and interfunction with parents or parental figure. This establishes an ongoing relationship with therapist as parental or other family figure, and patient as child/family member figure, at the financial cost of the patient. The patient is rescued from this professional exploitation of his predicament only upon his establishing an authentic, parental or familial surrogate–non-professional–or by his establishing an authentic, mutually nurturing relationship with mate, significant other, best friend, family or family member.

Characteristically, good therapy is often bad psychology: this is because the rejuvinated 'self' of the prevailing psychology schools, after some healing and strengthening through whatever is recovered of resiliency and of relationships, is still left constituted as an egoself, as the 'self' of 'developed' society and culture, and still remains in a largely unnatural, unhuman, unreal and untrue quasi life and quasi world with two tools, i.e., (a) the latest coping, conceptual structure and (b) the therapist as surrogate family figure, significant other or best friend. This arrangement provides for the continued security and growth of the psychological professions, serving their vested interest, and demonstrates the need for psychology to be grounded in philosophy–in a philosophy of mind and a philosophy of human being and reality–that is in turn interfunctional and unified with social philosophy and the other branches of philosophy as one body or vision of philosophy about the real, authentic nature of the human condition (of our subspecies). Human authenticism, the philosophy, fills this need.

## C. Alienism's Penetration and Fragmentation of Selflifeworld

Upon the penetration by alien, elements-phenomenona into natural, authentic, human selflifeworld, the whole, triecologic self is fragmented into bio-intellect-egoself beholding the other resultant fragments: diminished, disintegrated, adulterated family and community, and beholding the same damages to life and world; all of which now are dis-structured and dysfunctioning, following this shattering and adulteration of natural, authentic, organismic selflifeworld.

Fallen out of whole triecologic self, the diminished egoself feels insecure, unprotected and inadequate. To compensate for the miss-

ing portions of its subject–of itself–that is, family and community, and missing portions of disintegrated, organic lifeworld (including ecoregion), the egoself does three main things. One, it
bonds to an increased sense of sight for an inflated visual consciousness and scrutiny of objects and phenomena that are remnant pieces of original, whole unities, organics and organismicity.
Two, the ego bonds to altered senses and to diminished instinct
and intuition. To decrease its inadequacy, it values objects-phenomena that increase the import of the egoself to itself and other
egoselves–self esteem objects and social status objects. These are
attention-getting objects, primarily wealth and novelty objects.
Attention-getting objects serve to replace, though inadequately,
the natural nurture-care and sense of belonging–integrated, ecologic, harmonic identity–that had come with intergrowth of egoself, family-self, community-self, and organic lifeworld. Wealth
objects provide admiration-status and increased sensuality for increased egoself esteem. The same is true, with less consistancy, of
novelty objects–the new or the latest version; hence the ancient
birth and infancy of modernity, along with interjoined wealth,
both critical inflamations later in industrial socioculture. Thus,
novelty and affluence are newborn, emergent ethics of the separated, fallen egoself. The third main action of the egoself is to
bond to a swollen, transmuted intellect for the building of some
conceptual order of this now disorderly, transmuted self, quasi
egoself, quasi life and quasi world and to pursue power objects,
systems and phenomena that address directly its new insecurity
and vulnerability.

Unable to continue natural, human growth, the broken off
egoself joins with other egoselves to build the collective egoselves'
quasi life and quasi world. Human growth is replaced by human
success at this building of transmuted, quasi selflifeworld. Human
family-community growth to full, unified selflifeworld is replaced
by the building of career, organization, city, state, nation and/or
such like structures. Natural, authentic, human growth-fulfillment-
perpetuation is replaced by state expansion, the building of power-

wealth and its alien spirit-ethos, which is deleterious to the growth and perpetuation of natural, authentic human nature, being, spirit, identity and reality and their selflifeworld. Nevertheless, all that sci-technic 'civilization' has decayed of what is authentically human, human authenticism still continues to regrow with the resurging of The Last Human Spring.

The egoself is naturally and normally being held at the perimeter, outside of natural, authentic selflifeworld by the perennial, bequeathed perpetuation of intergrown bio-self, family-self and community-self until the instance of alien penetration splits four natural intergrowths: (a) of the whole self, (b) of whole mind, (c) of whole selflifeworld and (d) of whole Earthlifeworld. Though this occurs in an instant or perhaps actually simultaneously, our intellect, under the persuasion of modern science, tends to propose that within this phenomenon there may be an instantaneous sequence of these four fragmentations. Whether or not there is a sequence is important, but of secondary importance to the fact of the phenomenon itself, unless a determined first penetration can be judged as the independent, causal and preventable penetration within the general phenomenon of penetration; in which case it is of primary importance. This question has two components: the nature of the very first instance of alien penetration into natural, human authenticism; and the nature of all subsequent instances within each generation of humanity–when, how, and why in each generation of human beings. In this volume both simultaneous and sequential, sub-instance penetration are deemed worthy of consideration. At the same time, a first sub-instant within the instance or moment of penetration is offered up for consideration in Ch. V, "NATURAL, AUTHENTIC, HUMAN MIND."

The egoself, then, upon the instance of its emergence as fallen, transmuted entity commences a quasi life of beholding a quasi world predominantly as object(s). This is an *existence* fallen largely outside of human life and being, outside of all living nature; and suffering a large loss of being at that moment of bio-sociocultural mutation. Hence, the term existentialism, denoting a leading philosophy of, by and for this fallen egoself. The rebel, existentialist Heidegger would not see himself as an existentialist; for, he rejected life deficient in being by his very pursuit to recapture it and in his quest for the sense of being. He was on the road to authenticism, and in his later thought, at its gates, only to find that his egoself could not break through into natural, authentic, human selflifeworld nor the concept of it. (*See* "Heidegger" in Ch. X.) The loss is not just of human being but of natural being *per se*, a fall not just from human nature but from living nature. For, even the lower animals are beings within their organismic selflifeworld. Their forage, prey, shelter/home, water, air and Earth are not world objects but are lifeworld objects. Only sci-technic man largely and generally pursues observation of objects for observation itself and the 'knowledge'–data and information–of objects for knowledge itself. The sci-technic mind-set, perpective and consciousness misses the whole, wonderous, magnificent reality of selflifeworld, except on rare occasions; for, to fully acknowledge it would dethrown both science and personal careers.

The main part of the central dynamic is the attachment of eye to intellect to serve as intelligence for the fallen egoself. This bio-intellect-egoself, trapped outside selflifeworld and its being, beholds generally objects and their interactions. Observing (world) objects is a poor substitute for experiencing (lifeworld) objects. One compensation for the fall from selflifeworld into quasi intellect-egoself, quasi life and quasi world is to possess more objects and impart some life to them through status and the time and endeavor required to accumulate, utilize and/or study them. They are a part of us–an alien part–because they have taken a part of us–displacing part of our being, nature and reality.

The egoself cannot experience being; but, it subconsciously remembers it and has glimpses of it in the surrounding, still largely unpenetrated, unfallen whole, human selves intergrown with natural selflifeworld that starkly contrast its own makeshift, surrogate, quasi, bio-intellect-egoself, quasi life and quasi world. Feeling incomplete, inadequate, inferior, and unfulfilled while professing

superiority and success, this fallen 'superego' pursues excessive grati-
fication of the sense of sight for its fulfillment; as for security this
egoself adopts egoism and discovers that the 'knowledge' of objects that it is amassing is an instrument of power and its chief
weapon of defense and offense in its quasiworld of egoselves, forming alliances and building systems to compete with and secure
themselves against other such systems. And power is found equally
useful to subjugate, trick, bequile, propogandize or absorb the
multitudes surrounding their enclaved egoself systems, multitudes
that still harbor egoself's original competition and enemy: the force
of natural, whole selflifeworld, nurtureculture and its natural, authentic human being.

The egoself with still one root in biology has a natural base to
build an ethic of sight and sensualism into an ethic of objects–
objectism. But, two problems are immediately at hand: how to
make secure in the egoself a value for mere things–world objects;
and how to instill this in surrounding whole, human selves, still
untransmutated and unconverted, as a replacement for the diminishing or jeopardized, nourishing selflifeworld objects. Enter propaganda and later advertisement; the people can be taught to want
to possess and even less successfully to care about objects separated
from authentic, human meaning and purpose, from natural, authentic, human selflifeworld, its human being, nature, reality, and
the growth-fulfillment-perpetuation of these.

The rise of power–as opposed to family-community autonomy
and vested leadership–acts as the 'heart' of the quasi intellect-egoself
in quasi lifeworld. The ego-intellect has a 'feeling' for power. Prior
to transmutation, there is no power to speak of in the natural,
authentic, human society of natural, authentic, human selflifeworld;
there is only sufficiency-growth-fulfillment from the nurture-care
of family-community, a part of which is vested community leadership. The cynicism of 'civilization', the world of alienism, holds
that mother has power over infant, parents over children, family
head over family, clan head over clan, tribal chief(s) over tribe(s).
However, the egoself, having subjugated or exiled real, natural in-

telligence, has subjugated or dismissed bonded family and community, identity, and their nurture-care or love. It grows from and lives for sensuality, intellect and egoself power-wealth gratifications. Egoself power, an unnatural phenomenon, is the surrogate for the missing, natural nurture-care; for, in living nature and human nature, nurture-care and love are synonymous in diverse forms, diverse forms of living nature's Spirit. Nurture-care or love, when removed or subjugated, leaves egoself and its quasi life and quasi world without human being's central essence. Now without human soul, the fallen egoself has intellect as leader of mind, serving the senses, sightism, objectism and its own 'feelings' for power. Cynicism is the attitude for this 'super' egoself; power-wealth is its collective ethic and ethos. Egoselves relating to each other without soul, unable to love, devise therapy—concern-care—as a surrogate for the natural, authentic, human nurture-care of bonding, (of belonging with) family, community and intergrown selflifeworld and Earthlifeworld.

This egoself, which Freud partially glimpsed and inadequately portrayed as 'superego', replaces and fragments whole, human self. It has the intellect's 'feeling' for power and related data, facts and information, and the senses' increased, excessive sightism and objectism. In the hours, days and/or years of its activation, it rises from dormancy to domination and transmutes selflifeworld into bio-intellect-egoself quasi life and quasi world. It lives in order to function as bio-intellect-egoself seeking power-wealth and beholding, acquiring and studying things (objects); in order for which it must accomplish at least three tasks. First, to suppress the whole human self. Second, to build with other egoselves a super quasilife and quasiworld which is to dominate and triumph over natural, authentic, selflifeworld. Third, to build it to defend against and/or dominate other egoself quasilives and quasiworlds of different intellectual ideologies.

The egoself while trapped outside the dwelling realm of human being—human selflifeworld—is still welcomed there as a visitor. He speaks of his life of adventure-discovery, of its excitement and pleasures, not of its alienation, suffering and disillusionment. To tell the whole story would be to lose his image as hero and model. To admit to himself and others his loss of being would be to expose the masquerade, to deflate himself, his egoself; it would be to lose his new 'self', his inflated egoself; a social death at least and maybe death itself. He needs the people's verification that he and his breed are heroic 'supermen' to maintain this double self-delusion. These 'superpeople' form surrogate specialized quasi lives and quasi worlds they call clubs, associations and "communities", e.g., social, academic, scientific, governmental, bureaucratic, business, trade and professional. All things grow bigger and better, become 'super' like the 'super' egoselves that create them, including 'super' products, machines, weapons, homes, estates, cars, recreations, entertainments, corporations, nations, wars, mental and social maladies. All, that is, excepting four things which 'superself' neglects, dismantles and adulterates—natural, authentic, human, whole self, family, community and natural environment, essential sources of human being. Surely, our 'super' intellects, after all, will save us as they create their megacreation—'super' 'civilization'—unknowingly, the cemetery or alien jungle for their quasi, alienized species. But on the horizon of this darkness a light continues regathering, a warmth still resurges: the lifebeing-force of natural, human authenticism yeilds not.

## D. Individual and Social Arisings of Egoself

In addition to the historical, sociocultural moment of the emergence of transmutant, divergent, quasi egoself, it is equally vital to become aware of the many moments of the egoselve's emergence within each generation of 'civilization' socioculture. Within the blended, adulterated, nurtureculture of 'civilization' the whole self may experience a few or many fallings into egoself and separated, adulterated quasi life and quasi world. A falling may be brief, of longer term or permanent.

While the egoself emerges at several points within human generations of 'civilization', preeminently it emerges out of the alien, unauthentic, unnatural life period, adolescence. Adolescence doesn't occur in natural, authentic, human socioculture; instead, at puberty the breach to adulthood is made with a short passage period of a few days or weeks. Natural, authentic, human socioculture quickly and efficiently perpetuates itself simultaneously with biological perpetuation. In contrast, 'civilization' socioculture, especially 'developed' or 'First World' socioculture, prolongs the transitional passage from childhood to adulthood. Although triecologic selflifeworld in rural, partly industrial areas, largely blocks the aberrational emergence of adolescence and its deleterious, sociocultural transmutations, quasi intellect-egoself and its quasi life and world; this phony holding period becomes prolonged commensurate with the depth of 'civilization' into which one moves. Trapped in an artificial period outside triecologic selflifeworld, in a no-man's land (nonhuman)–between childhood and adulthood; the human self mutates in order to merely survive. Unable to become (grow into) adult, whole selflifeworld, the childhood selflifeworld mutates, falling out of selflifeworld into quasi intellect-egoself, quasi lifeworld. Thus, the human self is thereby divided. The egoself emerges to replace the whole, triecologic self. As fallen self, disenfranchised from natural, authentic humanity (from human triecologic selflifeworld), the egoself must build its own quasi life and quasi world.

With the intellectual awakening of 'civilization' 'mind' occurring shortly after puberty–generally one to three years, the matter of whether it occurrs in the unnatural, unauthentic period of adolescence, with the human self being transmuted into egoself, or in the natural, authentic state of adulthood, with the whole, adult self realized, makes all the difference: it is a matter of what is awakening–adult intelligence intergrown with adult mind and self, and lifeworld; or separated intellect employed by separated egoself. Remaining within whole, growing mind, intelligence continues growing, but separated from and overriding whole mind, emergent intellect commenses excessive, anomalous development in its employment by bio-egoself. It is, therefore, a matter of the nature of mind; and, therefore also, of human nature, being, identity and reality. Occurring in phony adolescence, with self as quasi egoself, the intellect's primary allegiance and service goes to egoself. Occurring in real adulthood this allegiance and service goes primarily to whole, human triecologic self. This switch within sociocultural evolution, of the general, predominating role of intellect, a switch in its belonging, from belonging intergrown and interfunctional with the rest of mind and to/with natural, authentic, human selflifeworld, to its belonging to egoself and quasi life and world is a primary moment of generational, nurturcultural mutation in man. And, since man is a social animal, it is a new kind of human mutation in two ways: it is instant, sociocultural, 'human' mutation; but, more precisely, since the change is caused by penetration by alien element-phenomenon, it is alien mutation.

On the generational level, adolescence is the major, commencement fall of man outside himself, of human nature in self violation. The pain of adolescence is that of being a freak, a freak not of biological nature, but of human social and sociocultural nature and sensing it. The near-proof is in the adoption of a separate appearance, dress and subculture. They know they are stranded between childhood and adulthood, and behave distinctly from both. Adolescents pay lip service to or compromise with the quasi-real, adult, unhuman, transmutant socioculture and its egoself, quasi life and quasi world until they are permitted to participate in it or conditionally surrender to it, which is only after years of institutional preparation, which includes indoctrinary education, general propaganda, seduction, bequilement, as well as, the media and advertising versions of these. However, something is lost and something changed forever; for, both growth and human growth is continuous, constant, timed, sequential and ingrediential. Growth put on hold is stunted and abnormal; human growth put on hold in a period of phony, unnatural, anomalous socioculture merely slows authentic, human growth-fulfillment while it is being intermixed with alien, artificial adulterants resulting in anomalous, transmuted, alienized growth of quasi, human, nature, spirit, being, identity and reality. Like plants pruned and trained to grow along strings or wires, people are irreversibly warped–and more complexly than plants.

Adolescense is the primary, episodic origin of egoself because, although the egoself usually substantially transmutes back into whole self upon mating and child rearing, it has by this time achieved complete formation and is waiting, suppressed inside, for conditions conducive to its re-emergence; conditions like the departure of grown children, divorce and separation from family, excessive career demands and numerous other penetrations of whole selflifeworld partially listed below. With its strong origin in the vigorous youthful beginning of biologic adulthood, the egoself's emergence over and replacement of whole self is by this advantage one of more frequency, of longer durations or even permanence.

In the particular cases of industrial and postindustrial societies a very common scenario of emerging egoself shows adolescense onset; followed or preceded by (a) a switch in primary identification with family and organic community–or what remains of these organic portions of whole human self left by the destructive elements-phenomena of alien socioculture–to peer group, to a city organization, to city and state; (b) adolescent adoption of regular non-addictive or semi-addictive drug use; (c) college attendance or envelopment in another egoself peer group or institution; (d) switch in primary identification with local organic community–or what remains if anything of this portion of whole self after alien socioculture's penetrations–to primary identification with an institution, career group, organization; (e) identification with city, state and/or nation over any remnant, small, organic community.

It is, therefore, when the pubescent youth lacks an adult selflifeworld at hand and a passage rite into it, because of the disintegration and adulteration of his and his parents' family and community, that the whole selflifeworld–of the child–cannot but break up into emergent egoself, quasi life and quasi world. The childhood selflifeworld has, tragically, often already been struck, shattered and transmuted. A notable account of what this volume calls alien penetrations into childhood is given by Joseph Chilton Pearce in: *Magical Child*. New York: E. P. Dutton, 1977.

However, it first must be noted that in preindustrial and industrial sociocultures of 'civilization', both the first, general penetration and the major, general penetration by alienism is the alien period of adolescence; however, at this writing, this may no longer hold true for post industrial sociocultures, where the number of penetrations and the magnitude of the adulterations from conception through birth through childhood are still growing of late.

The contemporary burgeoning of singledom in post industrial society is easily recognized as another major emergence of egoself; though neither being single *per se* nor the existence of a single segment in society is alien to natural, authentic human society. It is where and in what form, that determines this matter. Surely, both loss of spouse and ineligibility–unsuitableness–for marriage are natural and immemorial. However, both a rise in unsuitableness and a substantial rise in unmatchability of suitables are alien, and with a third alien, a large divorce rate, comprise all together a fourth, alien, sociocultural phenomenon alien to natural, authentic, human socioculture: a large, single segment emergent with post industrial, human socioculture. The first of these four arose in early, 'civilization' sociocultures which institutionalized ten to twenty percent into single living either through the judgement of unsuitability or through tradition in class related social scheme. The decline in the matching of suitables, on the other hand, generally is commensurate with city growth except where an effective, institutionalized mechanism, such as parental arrangement of offspring marriage, holds steady. Both burgeoned unsuitableness and unmatchableness are major deleterients in scitechno or "First World" socioculture and remain largely unrectified. Perhaps the major generating factor in unsuitableness is the progressive disintegration of family and community. With unmatchableness this disintegration is joined by the proliferations of specialized personality types carried along by the tide of general specialization of work and leisure activities.

Egoself, in fact, can emerge periodically or permanently during childhood, during adolescence and at several, main points within 'civilization' adulthood. These include prolonged, premarital, single adulthood (singledom); postmarital singledom (from divorce or death); estrangement from spouse and/or children; estrangement from family through heavy career demands; excessive demands of or absorption in career; the switch in primary allegiance and attention from family and community to career and/or the career peer group or some other peer group; impact by an excessive absorption in hobby, special interest or cause; the impact of regular ingestion of mind altering chemicals; the impact of *excessive consumption* of radio or television broadcasting, packaged music, books and periodicals; excessive consumption of entertainment, amusement, advertising, news, commercialism, consumerism itself; excessive idleness, activity, work, formal education, specialization, et al. The manifold activities of such emergent periods of fallen bio-intellect-egoself are responsible for the main structures of quasi life and quasi world built by the transmutant egoself. These egoselves are the designers and directors of this quasi lifeworld, and employ other egoselves and some bequiled whole selves in the building of the transmutant, quasi socioculture, unauthentic to our natural human nature, being and reality. All 'civilization' sociocultures are the design and creation of this transmuted egoself: the creations and behavior that are a conflicting blend of natural elements-phenomena authentic to us (Homo sapiens sapiens) and alien, artificial elements-phenomena.

The above partial list of transmuting, artificial, alien elements-phenomena at first seems like a list of excesses. Actually, once the egoself has emerged as a fallen, surrogate, alienized, unhuman egoself outside our species' elemental, authentic selflifeworld, and embracing intellect as intelligence, and embracing a quasi life and world; thereafter, any significant contact with these aliens triggers the fall, the breakup, the transmutation—psychic-bio-sociocultural-spiritual—of human being, nature, spirit, identity and reality. Thus, this is the original and ongoing source of the psychic dilemma which in the West has been generally termed, the 'divided self', among other terms; and which in the East has been perceived as the problem of getting outside the ego in order to experience human being. The latter perception was finally recently and notably duplicated in a Western version and approach by Heidegger in his later thought through his spiritual-ontological, existentialist phenomenology. The true source and notion of this widely recognized psychic dilemma within 'civilization' consciousness is ultimately revealed through the revelation of human authenticism as the alien, artificial transmutation of natural authentic selflifeworld into alien, artificial bio-intellect-egoself creating and engaging its adulterated, surrogated, fragmented, diminished quasilife and quasiworld.

Like the alcoholic who passes from being an incidental drinker to being an alcholic, a human transmutation has occurred, and awaits its next triggerings. The egoself, if not yet permanent, awaits suppressed, awaits its trigger penetrants that break the organismic unity of bio-self, family-self, community-self, natural life and natural world, forcing the ego into a repositioning of the remaining fragments into bio-intellect-egoself, which organizes with reason, science, philosophy, and religion the fragments of life and world–into quasi life and world–which had been, before fragmentation, organized as whole self, organismically, as intergrown selflifeworld. The egoself awaits inside for its preferred penetrant(s). Like a Mr. Hyde inside a Dr. Jekyl, it awaits for television, radio or music; for book, magazine or imagination; for pen/paper, paint/canvas or machine/project; for classroom, library or intellectual reflections; for singledom, single sex and single games; expensive cars, toys and hobbies; for novel excitements, sensations and sex; for novel discoveries, adventures and recreations; for new phenomena, products, and inventions; for another infusion of power, wealth or status; for more in quantity, size, and quality; and for other alien penetrants.

'Civilization' man's addiction–a true addiction–underlying both all other true addictions and mere habits of 'civilization' socioculture, is artifice itself–the artificial; the addiction is to the alien, artificial portion of 'civilization' within us. Once hooked, we cannot make it through the days, weeks and months of quasi life and world without the artificial, alien portion that is blended into the natural real portion. Incredibly, we are addicted to our own decayous, pollutant elements; because they are both expediently necessary (our expedient selection) and some are outright pleasurable. One might say that we have treated our decaying humanity with new various (opiate-like) pleasurable, artificial, elements-phenomena. The good news is that this addiction is, like others, withdrawable: through the light, goodness and joy of human authenticism rediscovered and re-embraced. Does not the alcoholic, by rediscovery and reembracement, start the journey back into whatever human reality, being, nature, and selflifeworld he can rediscover, recover and reconstitute?

It is from the revelation that family-self and community-self are true and real, natural and authentic portions of the natural, whole self authentic to our human subspecies, species, genus and family (and authentic as well to social primates and mammals), that the ego (bio-self) is revealed as only one natural portion of natural whole self; and that the transmuted egoself–bio-intellect-egoself–is revealed as a partly natural (biological), partly unnatural surrogate for natural, whole self. Moreover, since this blended 'self' consists of a reconstitution or reformation that has not occurred before–with the bio-ego incorporating inflated intellect as leader and nucleus of mind–this egoself must, therefore, be and is being tested by natural, human evolution for eligibility and fitness for incorporation into our natural, human subspecies and species purposes. Further, since this egoself incorporates into its quasi life and quasi world elements-phenomena that are outright artificial–outside of natural, human nature and being, and natural, human, organismic selflifeworld; it is therefore, doomed to fail its trial for incorporation into natural organismic, authentic, human being, nature, reality and their selflifeworld. We can be as sure of this as we can be sure that we cannot, ultimately, get away with pollution of the natural environment–and living nature.

Quasi egoself and quasi lifeworld, then, is being tested for organismicity by our organismic 50,000 year old subspecies through 'civilization' socioculture which is the general manifestation, creation, and realm of this alien, artificial quasi being. Since artificials are the opposites of naturals and since artificial opposites cannot interfunction as some natural opposites can, e.g., male and female, they cannot ever interfunction with natural organismicity: they can only intermingle or intermix as pollution-mixture–as adulteration and decay. This ongoing trial, ongoing error and failing experiment has spanned 5,000 years and occurred as the many episodes of diverse 'civilization' socioculture. The progressing decay in these sociocultures eventually reverses (sometimes suddenly, sometimes slowly) their development, moving them back into a recovering of the organismicity in human being, nature, reality and socioculture. Historians falsely and self-servingly record this cyclical phenomenon as a struggle of 'development' against 'undevelopment', of man-made, artificial, social order ('culture') against 'natural savagery'; or, more recently, as progress against some supposedly malign stability and continuity labeled by modern age thinking as 'stagnation'. But, how can a natural subspecies which is maintaining its natural being, nature and reality authentic to it, come to be judged 'stagnant'? Industrial man and his scholars, for the preservation of egoself, egoism, individualism, the idea of progress, and a sense of security, cultivate the illusion that self-driven historical forces that have engulfed him are actually initiated and controlled by industrial man. The reality is quite to the contrary: only the particulars of the general deviation and adulteration are under any significant and partial control. Historians, being among the few that benefit from church or state destruction, subjugation and oppression of natural, organismic, authentic a'civilization' (folk) socioculture, are disqualified at the outset as authentic, human truth sayers. Historians and scholars are the original propagandists and the original advertisers, employed down through the ages by various power-wealth ideologies. Traditionally, Earth-folk wisdom largely sees through the mascarade of 'civilization' while gradually being pulled into or conquered by it. Now, both those captured and those still free, have the human revelation in a written, complete system of philosophy, decribing and revealing our natural human condition, belonging to and authentic to us—the condition of human authenticism.

# CHAPTER V

# NATURAL, AUTHENTIC, HUMAN MIND

## A. Short Forward

THE PRESENTATION OF a particular, original philosophy of mind is a very complex task; there has been little in Western philosophy that seems to constitute a unified theory of mind *per se*. Rather, philosophy approaches the mind through the issues that are central to the philosophy of mind, wherever they arise in metaphysics, epistemology, aesthetics, ethics and particularly in the context of psychological philosophy and what is called the mind-body problem. In this volume presenting the philosophy of human authenticism, this traditional approach is used in regard to authenticism's philosophy of mind. Hence, the reader will find positions on the central issues to philosophy of mind scattered throughout this volume in its presentation of its metaphysics, epistemology, ethics, psychological and social philosophy and other branches, as well as in its portrayals of the nature of human consciousness, human being and nature, imagination, emotion, values, purpose, 'mind-body problem', egoself versus triecologic self, reason, intuition, senses, instinct, creativity and will.

Consequently, this chapter on philosophy of mind will not attempt a comprehensive sketch of human authenticism's philosophy of mind; but, instead will go toward emphasis, focus and elaboration. Therefore, it will also unavoidably restate some of what has already been stated elsewhere in this volume in different contexts concerning the philosophy of mind of human authenticism. And, to be sure, this volume holds that most of the problems addressed by philosophy of mind and by psychology arise from the illegitimate transmutation of whole selflifeworld into intellect-egoself, quasi life and quasi world, which is rejected throughout this volume by a major part of the human revelation, particularly in Ch. IV, EGOSELF, ORIGINS, DEVELOPMENTS AND ARISINGS and Parts C and D of Ch. III, THE REVELATION.

## B. Natural Mind from Natural Nutrients

Surely, it must be that natural mind functions through the natural feedings of it–through natural vision, other senses, and natural mental faculties, that is, with naturally integrated, natural senses, perception, intuition, instinct, revelation, and intellect organismically integrated as natural consciousness and intelligence, authentic to our 50,000 year old subspecies. After all, can natural, biologic bodies have healthy function and growth to maturity from a diet of part natural food and part artificial, alien nutrient! Surely not; and likewise, natural, socio-biological-spiritual-ontological human mind and natural, sociobiologic human being cannot have healthy function and growth to maturity from a blend of natural vision and alien vision, from a blend of natural intelligence and transmuted, artificial, alien intellect! Since, all life is of, through and with living nature, these unnaturals–these aliens and artificial aliens–blended into these living natures and into this natural, human mind and being, must surely produce pathogenesis and alienization. Microscopic and telescopic vision are dramatic and stark examples, equalled or surpassed perhaps only by television (*See* Ch. X, Part 36.) of how cultural invention can be not merely cultural transmutation but also human transmutation: transmutation of natural, authentic, human being, nature, spirit and reality.

## C. Transmutation of Mind and Intelligence

Upon alien, artificial penetrations and resultant transmutation, the separation of the egoself from whole, triecologic self is accompanied by the separation and transmutation of the intellect from whole mind and intelligence. The dislodged, transmuted, quasi intellect bonds to the ego, to the eye, to altered senses, and to diminished instinct and intuition, creating intellect-egoself; it discovers and studies constituents and focused images, i.e. objects now separated, fragmented and dysfunctioned from the organismicity of human subject-object–of selflifeworld. (*See* Ch. III, Part G, Selflifeworld as Subject-object.) Fallen from whole mind–from integrated senses, instinct, intuition, and intellect, and from whole self–the quasi intellect-egoself perceives, conceptualizes and builds a quasi world for itself of objects, facts, ideas and surrogative activities. Playing with these quasi objects-phenomena becomes the life of this childish, alien, artificial intellect-egoself, occasionally resulting in new, artificial constitutions of natural and/ or unnatural constituents. Two of these new, unnatural, alien objects, that are also artificial objects–the microscope and telescope– affixe to and interjoin with the natural, human eye producing alien, artificial vision, and, consequently, transmutant, alien, quasi 'intelligence' and transmutant, alien, quasi 'mind'. The microscope and the telescope furthered and ensured the flourishing and spread of the transmuted, alien egoself, by its effect of multiplying objects, i.e., an explosion of alien objects that constitute much of the quasi lifeworld of alien egoself. The mother of invention being necessity, the emergent and spreading alien intellect-egoself requires this expanding lifeworld; it must have a lifeworld surrogative to and competitive with, natural, human selflifeworld; it requires a lifeworld to embrace, engage, and attempt to reintegrate into a semi-functional mechanical system in place of a natural, fully-functional, organismic system.

An explosion of merely natural objects would have by itself been inbalancing. Tragically, this explosion of objects was one of *alien* objects and alien, artificial objects. For, while the microscope revealed the existence of many previously, unrevealed, constituent objects of living nature–natural, microconstituent objects and phenomena–the vision and knowledge of them was not by the natural human eye and mind respectively, since by nature the eye, senses and other mental faculties are interfunctional and intergrown. Thus, alien, artificial vision of them was not and cannot be intergrown and interfunctional with natural mind; instead, it is artificially interactive with transmuted, artificial, alien intellect-egoself. The shocking truth is that everything beheld by artificialized, alien vision is alien (to natural being and mind); and, all that knowledge derived from it is alien, artificial, unnatural, unreal, pseudo knowledge; furthermore, the alien mind and intellect that incorporate these two aliens (artificial vision and artificial, pseudo, quasi knowledge) into them increases their artificial alienism of quasi, alienized mind, intelligence, and being.

## D. The Household of Mind

The analogy of mind as a family household will be quite useful here. In a family its members use the household or home as a center of operations for their activities encompassing the community and the portions of their selves that are manifested as extended family-self, community-self and wider, natural lifeworld. The family household members interfunction together organismically in their constitutional creation of organismic, whole, human being–nature, spirit, identity and reality for each and all. This central or core realm of household members, in turn interfunctions with members of the community and/or extended family (sometimes as community) to constitutionally create an outer social realm (outside bio-ego-self or biological organism) of human nature, being, spirit and reality. Likewise analogously, this is the case with the faculties within the mind; instinct, intuition, intellect and the senses function together organismically to achieve an operations center for not just mind but for 'mind-body', since mind cannot exist without body and is interconnected with it through its contacts with body through senses. The so-called 'mind-body', the biologic organism actually, interacts organismically in the family household and in the more outward realm with other 'mind-bodies', i.e., those of immediate family, extended family, and community for the constitutional creation of natural, authentic, human selflifeworld, the medium, the structure-function-dwelling, of human being that encompasses, also, the interaction of these, their interactions with the natural objects-phenomena of a natural environment (of general living nature) and natural, human socioculture (nurtureculture).

Upon the penetration and transmutation of mind, natural intellect has alien objects-phenomena–perceptions, memories, ideas, desires, impulses, sensory data, et al.–in its room, making it what is now synthetic, quasi, transmuted intellect. Opposites do not attract and combine organismically in living nature and human nature like they do in inanimate nature; the exception being the realm of sex. It is when alien idea and objects-phenomena remain to participate inefficiently in understanding (intelligence) that synthetic, quasi intellect is created. The intellect is synthetic by virtue of its having synthetic ideology, not synthetic idea which doesn't exist, since the artificial and the natural cannot combine organically, but can only intermix. Natural perceptions, ideas and memories, et al., must now live with alien and alien artificial ones that are playing the part of these naturals which were exiled to the perimeter. These aliens seduce the intellect through an offering of power-wealth into accepting them as surrogates for the naturals rightfully belonging there. The natural, authentic, human mind knows that aliens are in the room of the intellect. This is an unprecedented, alien problem it is not equipped through natural, evolved, ecological design to handle. Until this moment of transmutation of natural intellect into unnatural, transmutant intellect, alien ideas and objects-phenomena have been exiled back to the perimeter, i.e., have not been granted places in natural intellect's room. The aliens are now, transmuted into fixtures, utensils, phenomena and ideology of mind's household family– visiting aliens become regular family fixtures.

Mind's household members, being equals by human nature, confront the now synthetic intellect about its transgression and transmutation from natural intellect into synthetic intellect, about its violation of the natural law, natural ecosystem and order of natural mind. Their outcry to adulterated, transmuted intellect: "A penetration and violation of one is that of all, for all function as members of the family of mind. Your self-violation of your family, of natural ideas and objects-phenomena, of natural ideology by the penetration of alien idea and object-phenomenon against house rules to re-exile them, is a violation of the natural order of natural, authentic human mind." The reply by synthetic intellect is a dec-laration of independence: "I have been called upon by the biologic egoself which has separated from the whole triecologic self, to form an alliance with it, the altered senses, the diminished instinct and the diminished intuition; and, our alliance shall henceforth serve as mind for the newly formed entity, bio-intellect-egoself; and, I shall serve as intelligence. You, my family members, will hence-forth be subordinated; and, I shall proceed to accept any and all aliens that gratify the new bio-intellect-egoself's need for power-wealth, curiosity (unnatural interest) and transmuted unnatural pleasure; and, we shall proceed to accept any and all of these new gratifications." This original violation of living nature, natural in-telligence and mind at the beginning of the dialectic of alienism is the moment of creation of partial, adulterated truth and reality, of partial, blended, human nature, being, spirit, and reality.

In the terms and view of idealism, it comes to this: synthetic or blended idea does not exist in natural, authentic human mind and reality; nor, contrary to the familiar, Hegelian thesis-antithesis/ synthesis model, does it exist, precisely speaking, in synthetic, blended 'civilization.' For, upon penetration and taking up lodging in the room of natural intellect, alien ideas or objects-phenomena intermingle with their natural counterparts, the result being a synthetic ideology within a synthetic intellect and a synthetic consciousness. It is the intellect and ideology that become synthetic as mixtures of natural and unnatural, alien objects-phenomena or ideas. Natural and alien ideas do not and cannot *combine* or interfunction—become organismic compounds of reality; but remain idea or object-phenomenon *mixtures*. There is no compound (synthesis) idea of natural idea and unnatural idea; there is only a mixture. Natural idea combines with natural idea; natural ideology with natural ideology within natural intelligence. Alien ideas combine to achieve alien ideology within synthetic, quasi intellect. Strictly speaking, there is no (purely) alien intellect for an intellect must be within some kind of earthly, organic mind; intergrown in this case with human senses, instinct and intuition. The only pure aliens that exist are transmuted alien, artificial, sociocultural elements-phenomena and alien ideas. However, the mixing of a natural and an artificial is also an alien phenomenon.

# E. Analogical Genetic and Mental Transmutation

Translated into the analogous context and language of genetics, this transmutation of mind means at least two basic things. Firstly, in biological evolution, deleterious mutant genes of the dominant type are exiled forthwith from the genetic pools of community and species. Secondly, deleterious recessive genes, when in the genetic pool, cannot manifest harm until they are joined by and combined with another recessive gene of their ilk, at which time the matched pair are forthwith exiled from participation. Nurtureculture mutation depicted in this volume is seen as parallel and analogical to that of genetic. A transmutant, alien element-phenomenon of behavior or community socioculture, wherein alien idea is one type of element-phenomenon, may be sufficiently disruptive to be exiled forthwith, or may be tolerated, as an alien, until it interjoins a key type of embryonic phenomenon–imagination–that has entered to visit. Their interjoinment attains them to transmutant, alien idea (Hence, the objecting expression: "the *idea* of such a thing!") and results in their exile. This is natural, authentic intelligence, wherein whole, organismic mind governs, recognizes and identifies transmutant, alien object-phenomenon exiling it from mind, nurtureculture, and selflifeworld.

However, following idealism's view and terms, perhaps it can be said that there happened along at the very start of 'civilization' a super, alien idea, the first and original one of its ilk, that penetrated and transmuted the natural mind and destroyed its power to identify and exile; that punctured the membrane of species identity, opening the way for the invasion of transmutant, alien objects-phenomena into our human species' selflifeworld. This artificial, alien, reconstituted, quasi mind, separated intellect and made it governing authority of mind, with other faculties being subordinated, oppressed, diminshed intuition, instinct, and altered senses; and ousted them from the chamber of power to a separate room. The first alien idea (following idealism's view) was the conceptualized picture of a new phenomenon–the separated, transmuted intellect-egoself beholding its lifeworld as separated unintegrated megaobject.

This transmutant egoself calls upon the intellect to study, make sense and use of two new super objects that it suddenly beholds. Mind, now transmuted, calls upon intellect as the faculty for this study, this studious endeavor. This transmutant mind of transmuted understanding–of intellect rather than intelligence–is incapable of beholding a natural, authentic (real) lifeworld, for such belong only to (intergrown with) natural, authentic whole self; it beholds instead intellect's first, original operation separation, i.e., the separation of life from world, as two quasi megaobjects-phenomena for study. It then begins the endless pursuit of separating or pulling apart each of these into constituents and their constituents into their constituents, toward the end of constituent understanding: ironically, constituents classified and organized by the intellect back together into groups and systems that will function as surrogates (organs) for the transmutant, alien, artificial intellect-egoself and its quasi social systems for its transmutant, alien quasi life and transmutant, alien quasi world.

Since, natural, authentic human growth and perpetuation, which provides the timed, sequential, ingrediential, orderly *change of growth*, have been replaced by transmutant, alien, artificial, power-wealth material and organizational development, which provides a *change of progressive diversity, disorder and progressive speed*; the end of alien, scientific, constituent understanding is inherently unachievable, because human reality, being, nature, identity and spirit are now in churning flux. Hence, each person, family, community, generation and socioculture are left with its pursuit of a partial human fulfillment. Even infected, decayous intellect-egoself, its quasi life, its quasi world and its transmutant, alien 'spirit' require a purpose if only in order to survive. For this transmutant, alienized, artificialized being, means and end are one: power-wealth emerges with its spirit-ethos.

Thus, the original, transmutant, alien idea delivered to transmutant, alien intellect-egoself from its transmutant, alien mind through the faculty of transmutant, alien intellect was, logically, a picture of, a representation of, an account of, an original transmuting, alien phenomenon: the separation and fragmentation of egoself from its other two intergrown parts, family-self, and community-self, and the separation and fragmentation of whole self from its intergrown lifeworld.

The analogy of the mind as a household is useful also in portraying the interactions of its faculties before and after alien transmutation. The mutation of mind is from whole mind, wherein the faculties are interfunctional and intergrown for the growth-fulfillment and perpetuation of natural, authentic, human spirit, being, nature and reality to transmuted bio-intellect-egoself wherein intellect commences periodic and sometimes lengthy or even permanent direction of the household for the development and maintenance of artificialized, dehumanized, denatured, alienized being and its power-wealth spirit-ethos.

## F. Fragmented and Alienized Dysfunction of Mind

In addition to the initial, primary moment of nurturecultural, nurturomic mutation–when the mind is penetrated and reconstituted by artificial, alien element-phenomenon, there are other vital moments of transmutation within the mind on our species' mutational road to decimation or extinction. After the intellect is separated from its natural role in mind and inflated, the senses undergo a progressive process of separation from each other, a disintegration of their organismic function and unity. Sight is separated from the other senses in the scrutiny of science, especially with the microscope and telescope and associated propositions, theories, perceptions and conceptions. The initial and accelerated mutilation by science of the senses leads the way into the general disintegrative mutilation of mind by sci-techno society. This nuturecultural, nurturomic mutation readily filters down into every day living. Our binoculars, telescopes and published photographs of microscopic and telescopic views present visions that, like those of any camera, cannot be touched, heard or smelled; and, unlike regular photographs, microscopic and telescopic photos cannot be given imagined touch, smell and sound. The radio and telephone present sound separated from the other senses. The telephone mutilates natural, human conversation and communication leaving little more than voices. The window at home and office reveals a section of exterior vision without smells, feel of breeze and temperature, and with virtually no sound. Television is the grandest, electronic, alien monster, surpassing its electronic, artificial relatives in presenting artificial sight and sound and goes further into mutilation in its presented images of directed, imagined, occurred and occurring scenes or news events by means of fixed, artificial lens or lenses (cameras) of fixed, selected, timed, aimed angles. (*See* Television and Authenticism in Ch. X.)

It is difficult to say, regarding the telescope and the microscope, which one has contributed more to the loss of natural, authentic perception, understanding, reality and being authentic to us. We lose sight of the forest about as easily from looking at tree bark cells as from gazing beyond them to the mountain tops and the heavens. It can be said of the man made artificial lens that anyway one looks through it, one loses, i.e., one ends up nearsighted or far-sighted. What a bitter paradox that this lens, first invented to correct faulty vision in particular individuals, now, further developed, is widely used by sci-techno humankind for a distorted and unnatural vision of its self, its lifeworld and its reality. For, nothing we see through either invention will ever save us from the untimate fear and insecurity we have reaped by looking through them. They were windows on Pandora's Box that eventually splintered the entire box (human realm) apart.

Through scopicized vision we split the atom and other particles of matter, physically and cognizantly creating a new fear and doubt that no defense, courage or faith can adequately absorb. The first task of human life is security from physical, mental or spiritual disability and/or death, i.e., the expectation of another day, another year, another few generations in the village of selflifeworld. Now, we have it not. The new deification of our age, science-technologism, or sci-technicism, has diminished human security, as well as human identity, being, reality and spirit. The surrogates for missing security become a regimated presence–hours and days of alien, artificialized work punctuated by novel, alien pleasures–and a future as projected, alien imaginings, fantasies and possibilities. Lacking the clear, certain, natural, real future, rightfully and authentically ours, we substitute the foggy, facinating, possible visions of existentialism-futurism–alienism. If we accept them we increase their degree of fulfillment–through self-fulfillment. Having surrendered our birthright security along with much of our birthright, being, spirit and reality, to our new divinity, science-technology, it is, henceforth, life one day at a time. Having surrendered our future, it is one game plan at a time. We are trans-

formed from natural children of our natural spiritual (and biological) source, Mother Earth–or of deified, natural elements of Her–
into subjugated, subservient quasi beings of science-technology.
Ralph Waldo Emerson's ominous warning has become progressively fulfilled prophecy: "Things are in the saddle, and ride mankind."

By trusting in the power of science-technology, we are assured
that life will go on and new glories will be at hand. We are supposedly to be delivered through this new alien power, and new glories, some unimaginable, will be achieved. By accepting the teachings of the new, 'science diciples'–scientists by the thousands–and
their 'ordained ministers'–engineers by the millions–we will presumably have new lives in the promised, brave, new world beyond
our grandest dreams, dreams from the new, quasi, clergy mind of
alienism (existentialism-futurism).

## G. Transmuted, Diminished, Partial Mind

The issue of mind within 'civilization' socioculture can be seen
also as the problem of partial mind. The emergence of partial mind
(as well as, partial being, spirit and reality) occurs with the emergence of 'civilization'. Since human consciousness is the central
phenomenon of mind, as well as, human being; partial consciousness is partial mind and partial being. Any partial consciousness or
awareness of a mind's–and being's–lifeworld, i.e., of its selflifeworld,
is consequently also, diminished, partial perception, intelligence,
life, being and spirit, just as it is partial consciousness and self-
consciousness. A main problem of sci-techno man is his having
only partial mind and consciousness. To have whole mind and
consciousness one must engage whole mind (and consciousness)
with natural, authentic, human, whole self and whole lifeworld as
human selflifeworld, within Earthlifeworld (ecosphere). One falls
short of full mind function while mentally engaging natural phenomena disproportionately perceived, as well as, while engaging
the many distinctive, artificial, alien phenomena of sci-techno

socioculture. Two examples, respectively, are: (a) the engaging of a natural phenomenon or phenomena in excess or in deficiency of the proportion it or they constitute in natural, authentic, whole, human selflifeworld consciousness; and, (b) the engaging of quasi, artificial, alien phenomena–adulterated, fragmented, surrogated, diluted phenomena of artificial, alien, 'civilization' socioculture.

To set about to rediscover and recover–what real, whole, human mnd and consciousness is–what is natural mind, authentic to our 50,000 year old subspecies–is to go looking for not just the unadulterated, untransmuted faculties of human mind, but is, at the same time, to go looking for the unadulterated, untransmuted objects-phenomena that the pristine, human mind was conscious of–what filled its consciousness. The two will be found together unseparated by artificial penetration. And, it is likewise for the recovery of other losses, like partially lost human reality, being and spirit.

Beyond the tragic and ugly plight that 'civilization' man is loosing his mind is the nature of this diminishment: the mutilation of mind. Sci-techno man is undergoing an ongoing, generation after generation functional lobotomy, with variations and additions being made with each generation, and with each year for that matter. What is happening concomitantly to the biological brain of sci-techno man, aside from organic based pathologies of some mental diseases, could be partially discerned by unbiased use of brain test technology applied to each of the two groups, i.e., especially by comparing the PhD brains to those of their nearest counterparts in a'civilization' Earth-folk sociocultures. But, where to find unbiased scientists without the biased presupposition that any results are to be interpreted as indications of their own 'superior' educated brains?

# H. Transmuted, Diminished, Artificialized, Alienized Being

We have some inherent capacity to deal with diminishing, material resources–scarcity–as there have normally been lean years for communities and sociocultures due to drought, game shortage and crop failure, and so forth; such are temporary, material short-ages from these recurrent, natural phenomena. However, we have no inherent means beyond natural, human surrogacy and substi-tution to deal with diminished, human nature, being, spirit and reality. Thus, when these drop below our species requirement–become deficient–the egoself is born: the intellect separates from its family of natural mind and bonds to the ego, inflated sight, other altered senses, diminished instinct and intuition. This intel-lect-egoself is incompetent here because it can only engage in the understanding of constituents and of the fragments and reorgani-zations of them left after the big bang within the self when the mind was struck by artificial, alien element-phenomenon, thereby, reconstituting it as transmutant mind. At our essence, the spirit of Homo sapiens sapiens could not withstand the impact upon the mind by this alien element-phenomenon; and, resultingly, intellegence transmuted downward into intellect, and human spirit and being transmuted down into an alien, quasi spirit and being. Spirit must have mind; and mind, spirit. The transmutant mind and self, therefore, receive a reconstituted, transmuted, quasi 'spirit'; they receive the 'spirit' within power-wealth spirit-ethos. The four mutants (mutant mind, self, spirit and being) form a united, mechanistically, rationalistically (intellectually), integrated one, Homo sapiens alienus. This new quasi entity is not a new, *natural, real subspecies* but is partially real and partially that portion of our own natural, human subspecies which is decaying from the patho-genesis, alienism. It is doomed to extinction without the diagnosis and treatment springing from the human revelation. It is partly outside living nature and human nature and being; having a self, a mind, a lifeworld, a being, and a reality, that are part natural, and part artificial and alien; and having a 'spirit' that is actually wholly alien.

This new, quasi human, from the beginning commences at least two, unnatural tasks necessary to continue outside of natural, human evolution. First, it must build, maintain and reproduce with each generation a quasi self, quasi life and quasi world that, being generated by the dialectic and pathogenesis of alienism, is progressively artificial and alien. Second, the intellect at the lead of transmutant, alien mind is incompetent here, i.e., it can only 'understand' constituents and its reorganized systems of these, not their ecologic, organismic interplay within human reality, being, nature and spirit—nor the interplay of these with Earthlifeworld.

# I. Organismicity of Mind, Intelligence and Will

The natural mind of any natural species uses whichever faculty it requires for natural intelligence (efficacious, naturally evolved, ecological living). How does a species know which is required for different situations? *It* doesn't know: the mind instincly responds correctly; and, then, afterwards, the bio-socio-spiritual-ontological organism (triecologic self) may or may not be *aware* which faculty was used. It is the whole, triecologic self that *knows* through conscious and unconscious intelligence how to correctly respond. Further, this intelligence is interfunctional and intergrown with and aided by natural, human purpose (or will) which wills toward the growth-fulfillment and perpetuation of human species' nature, being, spirit, reality and the structure-function-dwelling—selflifeworld—of these.

There are natural, gender roles in natural, a'civilization' Earthfolk socioculture, for example, with men hunting and women cooking and child rearing; but the result is only a slight, significant difference in the male and female balance of the mental faculties. The external differences are matched or fitted to internal ones—to mental, social and sociocultural differences. But, this slight difference can be substantially increased or decreased by alienized, alien, 'civilizaiton' sociocultures.

The natural, authentic, human mind has no nor needs any directing or leading faculty. The faculty best fitted to the situation or phenomenon responds; in the same way that any one sense may on occasion lead or predominiate the other senses. In both cases all the agents or faculties are not functioning on behalf of (organic to) the mind *per se* nor the mind-body (organism) *per se*, but rather are organic to (and functioning on behalf of) the organismicity of human being, nature, spirit and reality and their selflifeworld, which is self-directing through species' selflifeworld purposes/intelligence. Purpose (will) and intelligence, in one sense, are two parts of one faculty! More fully, they function with spirit as life-being force.

They are one intergrown, interfunctioning phenomena, within consciousness and subconsciousness, of natural animal and human-animal reality, being, spirit and nature before the transmuted intellect-egoself separates them and all the above constituents—mental, physical, social, spiritual, material, spiritual-ontological—toward the goal of constituent study or understanding, classification and reorganization for the fallen, adulterated, diminished egoself, its quasi life, its quasi world and its power-wealth spirit-ethos.

How can intelligence and purpose (including will as conscious purpose) and spirit be three parts of one thing! They are because they are two interfunctional, intergrown parts of a whole, *complete* faculty. After all, can a mother with a natural purpose of nurturing the growth of her children do this competently with *retarded* intelligence and diminished spirit and being! And, does an idiot have the normal, human purposes, goals and vital skills toward fulfillment and perpetuation!

## J. Natural Perception Vs. Transmuted Perception

The problem is that alien objects-phenomena enter the mind and cannot be perceived, cannot combine into or interfunction with the natural elements and faculties of natural mind and consciousness; because natural perception is interfunctional and intergrown with natural mind which in turn is interfunctional with natural being, reality and spirit. An unnatural, artificial adulterant cannot join into–cannot enter into interfunction and intergrowth with–natural being or any parts of it such as mind and perception. The artificial, alien objects-phenomena, then only intermix as pollutants–instead of combining with interfunctional growth as participants; they only diminish and decay the natural growth and perpetuation of natural being and all its natural parts, from perception right on through natural socioculture (nurtureculture). Thus, alien, artificial objects-phenomena enter-penetrate–perception but cannot enter into perception's capacity and function–into its natural function of interfunctional growth with ecologically integrated intuition, senses, instinct and intellect; the result is an adulterated, quasi form of perception: conception wherein intellect separates from integrated perception, inflates and repositions into dominance over intuition, senses and instinct. This transmutant–conception–suppresses natural perception and consciousness; just as its transmutant manifestation–intellect–suppresses the manifestation of perception and consciousness–intelligence.

To close on the subject of the nurtureculture transmutation of natural, human mind, consciousness, intelligence and being; two parallels are to be sighted that are analogous to and supportive of athenticism, its concepts concerning human mind and its dialectic. Firstly, contemporary science holds that the human body has, at all times or at many times, deleterious bacteria, viruses and cancer cells, which its immunological system holds to tiny numbers incapable of participation in and disruption of organismic function. We know that once they overthrow this defense system, these aliens–not belonging to/with organismic health–will within a period of days or years progressively destroy the body and the person. Secondly and further, it is obvious that every species has a genetic mechanism similar to the immunological mechanism for maintaining its nature, being and reality. From these two, it is tenable and

*even logical that social species, having evolved ever so slowly an exten-*
*sive nurturecultural, nurturomic feature and realm of their nature, be-*
*ing, identity and reality, possess not only a genetic system and mecha-*
*nism for growth, self-perpetuation, and species defense of genetically*
*based nature, being, identity and reality: social species also possess an*
*additional system and mechanism for growth, self-perpetuation and*
*defense of species' socio-psycho-spiritual-ontological, nurturecultural,*
*nurturomic nature, being, identity, spirit, and reality. The second natural*
*realm and its defense system is the missing, second half to the first,*
*genetic realm of living nature: it is the missing link, linking human*
*culture to evolutional life as a partner with genetic nature in life's evo-*
*lution.*

# CHAPTER VI

# DIALECTICS

## A. Short Forward

THE PROSPECT THAT 'civilization'-humanity is caught up in a dynamic of adulteration and decay (disguised as "development"), instead of a dynamic or process of growth-fulfillment-perpetuation, would be too depressing to be pragmaticaly tenable, if it were not a revelation that comes with the revelation of the authentic, human dynamic, the lifebeing-force, of natural, human growth-fulfillment and perpetuation. Through cultivation of our natural, authentic human dynamic, those imprisoned within the historical dynamic can achieve substantial releasement, and thereby commence the recovery of what has been lost, the cleansing of what has been adulterated and the unity of what has been fragmented of natural, authentic, human nature, being, spirit, identity, reality and selflifeworld. The unthinkable becomes thinkable with the newly beheld, achievable redemption–the awareness of and the heartening participation in the Last Human Spring.

## B. Organismic, Human Reality

Human reality is human being as natural, organismic objects-phenomena, participating in natural, human genes, nurtureculture, and selflifeworld; all in turn, participating in natural Earthlifeworld.

Indeed, living reality entails *organismic contituents*. And, living human reality is organismic, constituent, nature, being, spirit and identity. A constituent, therefore, is outside living reality until it is accepted into an organism. Hence, in this fuller and more real sense, a human gene is not yet real: that is to say, it is not a part of living reality until it is accepted into and belongs to/with a living organism, until it participates in the living phenomenon of organismicity. Likewise, a transmutant, alien nurturecultural element-phenomenon held at the perimeter is not participating in any organismic phenomena and is not yet a part of reality. When, in living nature a mutant gene enters through acceptance, into living organismic phenomena, it enters living reality. Likewise, within living, human nature, being, reality, selflifeworld and nurtureculture, a mutant gene or transmutant, alien, nurturcultural element-phenomenon enters living organismic phenomena, it also enters and participates in living, organismic, human being-phenomena.

Incompatible, alien entrants are exiled back to the human perimeter, having been tested and found unsuitable to participate in living organismic phenemena. Just as, in the sphere of the organismic phenomena of living species, these entrants are judged on suitability for living, species, organismic phenomena; likewise, in the sphere of highly evolved, human, organismic, living phenomena, the judgement on suitability is for participation in the reality of human species nature, being, spirit, identity and reality. In natural human, evolution, human change or novelty—as transmutant alien objects-phenomena—must await at the perimeter for acceptance into the natural ecosystem of natural, authentic, triecologic selflifeworld. Perspective, entrant elements must be judged participative in the individual self's growth into acquiring interfunction with family-self and community-self, as well as, that matured, whole, triecologic self's participation in its natural, authentic lifeworld. This dynamic of human lifebeing-force, as just described of human reality, is penetrated, violated and adulterated by the dynamic of alienism–by quasi, alienized, artificialized, human unreality. A dialectical struggle then begins.

## C. Living Nature; Natural, Authentic, Humanity, and 'Civilization'

Human authenticism, the philosophy, affirms and describes the lifebeing-force within natural, authentic, modern humankind (Homo sapiens sapiens)–the force moving the human being and reality, the human nature and spirit, that is authentic to our approximately 50,000 year old human subspecies. Our natural, human lifebeing-force, as well as, our natural, authentic, nurtureculture and selflifeworld have been in a dialectical struggle with the distinctive, alien forces and phenomena of artificialized, 'civilization' societies.

Living nature has her own lifebeing-force, life-force and reality, generally recognized as natural selection and mutation–evolution. And, therefore, living nature of which we are a part has also been engaged in her struggle with 'civilization's' alien forces and phenomena suffering increasingly heavy damages ("environmental" or "ecological") during the twentieth century.

For humanity, the dialectic between our human lifebeing-force and the artificial, alien forces within 'civilization', precedes and envelops smaller, minor dialectics, such as the Hegelian dialectic. This major dialectic, the dialectic between human reality, being nature, spirit, selflifeworld and the alienizing, artificializing, adulterating elements-phenomena of 'civilization' is the dialectic of human authenticism versus alienism. This dialectic and the human consequenses are revealed in this volume with the revelation of the philosophy of human authenticism.

This human authenticism vs. alienism dialectic, humanity's basic dilemma and problem, was preceded by the primordial, dialectical struggle of living nature and her human nature and being against the forces that would reverse species evolution and all evolving life back toward extinction/death, or inanimate matter. In both dialectics, 'thesis' is opposed by 'antithesis', but eternally defeats 'antithesis', taking from it any beneficial fragments that it occasionally finds. There is no genuine 'synthesis', no new genuine 'thesis', in either dialectic; there is only reigning evolving 'thesis': living nature in the first, natural human nature and being in the second. Their 'antithesis' (opposites)–(a) unliving nature, and (b) unhuman, denatured, alienized, quasi, human being and reality, respectively,—contrary to appearance and popular presumption, cannot be and are not incorporated into their 'theses', i.e., into living nature, and human nature and being. Only a tiny percentage of mutation is accepted into these two theses–living nature and human nature–to benefit and belong to/with her. This does not constitute a general reconciliation of opposing 'theses'. For, living nature and human nature each embrace themselves; they do not merge or combine with unliving natures to produce true 'syntheses' or hybrids: they only intermix to produce (a) quasi living nature– the polluted ecosphere–and (b) quasi, artificialized, alienized human being and reality. It is only 'civilization' man that intermixes and merges with elements of unhuman denatured, alien, and alien being, acting through his violated, transmuted will, nature, and being. As for living nature, she produces her own 'alien'–exiled nature–confined mutant genes–to remain in indefinite exile at her perimeters until a tiny portion of these mutations can benefit her, taking them at that time into her bosom and ecosystem, to join 'thesis'–to become natural. Incrementally utilizing her stored adaptables (mutant genes), she knows and embraces her nature, identity, being and reality, in the same way as she knows and embraces life through reproduction. Using mutation, she continues her self knowledge–of her identity, her nature and diverse species being.

The second dialectic has become our paramount, most crucial human drama: natural human being versus alienism. Here, human nature, being, spirit and reality struggle against adulterating, artificializing, alienizing elements-phenomena. This dialectic has incrementally intensified with the incrementally increasing artificialization and alienization of industrialization during the nineteenth and twentieth centuries. Here, the 'thesis' of natural, authentic humanity is opposed by the 'antithesis' of unnatural, adulterated, transmuted, alienized, quasi humanity. In this 5,000 year old, non-triadic dialectic, the opponents struggle continually much like the growth of spring and summer with the decline and dormancy of fall and winter, and much like health versus desease. Though human nature and being, and their human lifebeing-force have in the long run been losing this struggle, there have been times, places and epochs in which human nature and being regenerate and experience substantial or general recovery: humanity, as well as particular communities, families, persons and societies, experiences such human springs.

The incremental progressive opposition to humanity's natural, authentic nature, being, spirit, reality and selflifeworld by 'civilization's' adulterations, starting with the first cities, is the original, human 'thesis-antithesis'. For over 40,000 years our subspecies, modern humankind, was of living nature, a member species within a family of species progeny, among living nature's many other species progeny. The departure of humanity from its natural authentic home, its selflifeworld (*See* "The Revelation of Human Selflifeworld," Part C of Chapter III.), human, whole self, organic family and community, human nurtureculture (*See* "The Revelation of Nurtureculture and Nurturome," Part 10 of Chapter III.), into transmuted, artificialized, alienized surrogates of these, is the original, human dialectic–the humanism/alienism dialectic–wherein natural, authentic human nature and being, and quasi, alienized, human nature and being are in oppositional struggle. This new, dialectical phenomenon is 'civilization' emerging within and opposing natural, authentic, human nurtureculture and selflifeworld. In it, artificial, alien adulterants oppose and thwart the natural, authentic, human process of growth-fulfillment-perpetuation, and grows within our human being and reality like cancerous growth within healthy growth.

The first two dialectics–(a) living nature's evolving life versus regression and (b) human nature and being versus alienism–are not triadic dialectics of the Hegelian triadic form, wherein 'thesis' is opposed by its opposite, 'antithesis', merging with reconciliation into 'synthesis'. For, a mutant gene never merges with a normal gene to create a reconciled gene; it merely replaces it after waiting indefinite, uncertain chance to fit into one of living nature's ecosystems. Likewise, mutant, human species' nurtureculture cannot reconcile with natural, authentic, human species' nurtureculture, but instead is kept outside or on the perimeter of selflifeworld by family-community law, by the forbidden and taboo. When an element of alien, anomalous nurtureculture–a behavioral element or social element–breaks into natural selflifeworld, it is met with sufficient rejection to deter and exile it back to the perimeter. Since the triecologic self of self/family-community needs acceptance and identity with all three of its interjoined, interfunctioning portions their unity assures that these breakins remain incidental and nonproblematic. Only when the sovereign autonomy of natural, authentic, human selflifeworld is generally and continually violated, and when invaders become occupants–when human nature, spirit, being, identity and reality are transgressed by historical force, by the opposition of 'civilization': it is then that the reigning, self-perpetuating, authentic dynamic, the lifebeing-force, of human nature, being, reality and selflifeworld ends and the dialectic of humanism versus alienism begins.

Until this penetration, the natural, authentic, human dynamic or lifebeing-force of human authenticism, had kept transmutant, alien nurtureculture at the perimeter of natural, authentic nurtureculture.

(It should be mentioned briefly here that authenticism holds Hegel's triadic dialectic to be a dialectic of only one of the thought processes of 'civilization' humanity. It does not apply to all 'civilization' thought; it does not apply at all as a dialectic of natural, authentic, human thought. Moreover, it is not a dialectic of life and reality, nor human life and reality.)

This breakin and violation of the natural, authentic nature, being, spirit and reality of our species perhaps occurs through the membrane of identity. It is the intellect that rebelliously and falsely judges transmutant, alien elements as acceptable and worthy to belong and participate; doing so as it aligns with and serves the egoself, which has separated and fallen from its other two parts of whole self, i.e., family and community (*See* "The Revelation of Triecologic, Whole Self", Part 15 of Chapter III.). This seemingly is the instant of social and nurturecultural transmutation; when the ego falls from whole self and adopts intellect, altered senses, diminished instinct and intuition as mind to replace natural, authentic, whole mind, i.e., natural intellect, senses, instinct and intuition. In this instant also, intellect falls out of interjoinment with spirit, heart and being. The transmuted intellect-egoself beholds its new, quasi selflifeworld with the body and quasi mind as self, and power-wealth objects, systems and phenomena as quasi life/world. Hence, the biblical admonition, "for what should it profit a man if he shall gain the whole world and lose his own soul?" (*See The New Testament*, Mark 8.13.).

Whereas the original 'antithesis' (opposite) to living nature is the unneeded, harmful portion of her own mutation, mutant genes, created and held as available options; the original and only 'antithesis' to human nature and being is dual transmutation; for, (a) genetic mutation is later joined by (b) social, nurturecultural mutation in the human species having a social species nature. Living nature incrementally selects from mutation that which benefits and perpetuates her; she selectively incorporates mutant genes into new compositions of her. In our 50,000 year old human subspecies, genetic mutation is of secondary importance, for two reasons. First, in humans a generation is 20 years rather than weeks or months as in lower species of life. Second, and more important, human beings are highly social animals and the most culturalized. This socioculturality progressed to 'civilization' whereupon it gave rise to a new opposition to living nature, a new 'antithesis' occurring in the species, Homo sapiens, that overrides genetic mutation. This most socio-culturalized of animals, after about 45,000 years of social, subspecies nature, which followed upon perhaps

250,000 years of our species nature, and over a million years of socio-cultural human family's nature (depending on one's selected time of human origin), crossed then a new threshhold and experienced nurtureculture mutation. The social and cultural species nature of modern humankind was penetrated by the combined force of at least two interacting forces; two and perhaps more creative constituents combined to form a new megaforce phenomenon. After perhaps a billion years of living nature, and about 45,000 years of social, subspecies, human nature, a new opposition to living nature was created by the unnatural interaction of (a) the unnatural force of artificially created, unchecked overpopulation and (b) the unnatural force of the spirit-ethic and spirit-ethos of power wealth.

The difference between the dynamic of alienism and the one it has begun to work within and against, the dynamic of human authenticism, is that the latter is the growth and perpetuation of modern humanities nature, being, identity and reality (thesis) reigning perennially over opposition (antithesis). While the dynamic of alienism is the dilution, adulteration, and pathogenesis of natural, authentic humankind as thesis through the progressive, successful penetration by its antithesis, i.e., transmutant, alien, unhuman, denatured, alien being, unidentity and unreality into a basically and ultimately false, unviable reconciliation of the two: synthetic, blended, quasi humankind. Hence, the original, authentic, human dynamic, wherein thesis reigns and does not reconcile with antithesis (confining it in exile) was penetrated at and within 'civilization' by alien elements-phenomena creating a false deleterious, triadic dialectic, unsustainable in life's evolution.

The difference also is that the dynamic of alienism grows and swirls within the dynamic of human authenticism, within natural, authentic, human growth, fulfillment and perpetuation; this is the growth of human decay within human growth-fulfillment-perpetuation. Human pathogenesis has been recognized and resisted in the biologic, mental and social spheres of humankind. Authenicism recognizes and resists this decay in the whole and

unified sphere of human nature, being, identity, spirit, and reality. Authenticism as the growth and perpetuation of whole, unified, modern humankind has been, since its penetration by 'civilization', in general, progressive–though cyclical–decline in its
struggle with alienism. But now, human authenticism has a philosophy that reveals its dynamic and the dynamic engaged in mortal opposition to it, alienism. Hence now, we have entered the
epoch of The Last Human Spring–of the recovery, unity and preservation of natural, authentic humanity, its selflifeworld and
nurtureculture

The dynamic of human authenticism evolves humanity,
through incremental, natural selection from stored away genetic
and nurturecultural mutations, in a growth and perpetuation of
modern humanity. The dynamic of alienism transmutes humanity of 'civilization' through accelerating, unnatural selection and
accumulation of nurturecultural transmutation, into a progressive
progressively, illusionary human nature and being–a progressive,
human adulteration synthesis. This synthesis does not reproduce;
it dead ends in sterility. It ends at pure synthesis–pure, artificial
human nature and being, which does not exist; it ends at total
elimination of human nature and being, at elimination of humanity. Only nature and being can reproduce–can be evolutional self-
perpetuation. This synthesis either increases or decreases–more
precisely, it either insynthesisizes toward unhuman denatured, alien
being; or it desynthesisizes toward natural, authentic human nature and being. Desynthesis (of human authenticism) ends at recovered and unified natural, authentic modern humanity;
insynthesis (of alienism) ends at unhuman unliving nature-alien
being, which is outside both living nature and living being: it
ends at elimination of humanity. It can biologically reproduce
because it has sex from its biologic portion; but, it is quasi reproduction–it reproduces only its natural portion. The alien portion
of synthesis cannot experience natural reproduction, since alien–
and artificial–being cannot have natural reproduction. Henceforth,
each generation inherits part of its nurtureculture as natural, and

part as alien. One alien element is artificial change–nonreproduction and nonreplication; and the quasi intellect, now synthetic intelligence, is charged with copying some alien, quasi, false 'nurtureculture' and innovating the rest of that which artificial change–nonreproduction and nonperpetuation–has not reproduced. Natural, authentic human change is only that which is within the natural, authentic dynamic–the selection of incremental genes and nurturecultural elements-phenomena being created and held at perimeter, which are beneficial to the growth-fulfillment-perpetuation of human species' selflifeworld. All other change–selected by the quasi intellect-egoselves–brings harm (often hidden) that outweights benefit. Species natural genetic and nurturecultural change that is required to maintain its fit into living nature's ecologic system perpetuates the species; all other change is deleterious and movement toward decimation or extinction.

# CHAPTER VII

# THE BIG BANG WITHIN EVOLVING LIFE: ORIGIN AND EMERGENCE OF ALIENISM

## A. Ground Zero: 'Civilized' Humankind

EVOLUTIONARY THEORY OF evolved life is incomplete without a cogent explanation of life-destructive 'civilization', a phenomenon that clearly opposes evolutionary life and its ecosystems of Earth's evolved lifeworld. We need to realize that at the core of 'civilization' lies an embedded, degenerating force alien to all Earth-life, a force-phenomenon that debilitates and commonly reverses the evolution of life through its pathogenic debilitation of natural human being, spirit, and selflifeworld.

Alienism is the unprecedented—in four billion years of evolved life—alien phenomenon of explosively increasing anomalies and alienations attained to transmutation, a process destructively opposed to life, its lifebeing-force and its ecological, organismic world (field) of beings, including human being. Alienism is manifested by the intellect-egoself which has usurped the power to have its desires (wants) subjugate the legitimate, essential needs of the natural, whole, human self of whole human being within living nature's world of beings, Earthlifeworld (ecosphere).

The exponential explosion of novelty is nothing less than the ongoing, destructive explosion of human nurtureculture and whole selflifeworld. It is unprecedented in our three million-year-old, evolved human nature and being, and in four billion-year-old, living nature herself. This novelty explosion entails that our human being experiencing this explosion within its selflifeworld is being alienized exponentially out of Earth's evolved, natural human being. This is so, because modern novelty–and that of 'civilization' itself–is distinctively different and beyond the genetic and nurturecultural, mutant 'novelties' appearing, circulating and cycling within our species. Any similar explosion–known or supposed likely to have occurred in the universe–would still be alien to Earth's evolved, living nature.

There are many artificial things and phenomena in our 'civilized', quasi, alienized selflifeworld. Since the many, natural, species-beings of living nature would not make or do artificial things that are outside the nature of evolved life, these must have been created or initiated by other beings–they must have been created or emergent from human being that, by virtue of such artificial behavior, are no longer purely and wholly within natural, human being (beingness). Alien, artificial objects-phenomena must be emerging from a part of our human being–from a part of our species human being and beingness. These artificial objects and phenomena emergent in human selflifeworld are alien by virtue of their inability to participate in living nature's systems without destroying–in an unprecedented manner to an unprecedented degree–vital parts of her natural systems. The existence of objects and phenomena that destroy parts of living nature and human nature progressively diminishing the depth and breadth of her evolved, organismic, living being, of her evolutionary system of life, and our, evolved human being, is proof beyond an intelligent doubt (reason is a defense and rationalization of 'civilization') that these artificials are also aliens;

and that alien being emergent within our, natural human being must have created them.

Living nature–evolutionary life–is penetrated from outer space
by alien comets and meteors. Once, sixty million years ago, one
was so large as to alter and reverse the evolution of Earth-life. Living nature required eons of time to treat this alien presence or
pathogenesis in her system; after a million or more years the decline in her life and life systems was reversed and evolutionary life,
via renaturalization, recovered the normal nature and process of
Earth's evolutionary life. Eons later Mother Nature had recovered
from this debilitating violation. The aliens of 'civilization' also penetrate from outside evolutionary life; but, the accelerative impact-
explosion is from within living nature's human species. The decline in Mother Nature's life is not from an instantaneous impact:
rather, living nature has struggled against this recent, unprecedented
emergence of alienism for five to ten thousand years in cycles of
human alienization and renaturalization–as seen in cyclical 'civilization'. Can evolved life recover from or arrest this new,
unprecedented alienism? Indeed but the question remains: will
recovered Earthlife salvage the human species. The answer lies
within us, within human nature's limited ability to diagnose and
treat this alienism in living nature's human being. Some human
communities will have, indeed, received the human revelation and
will possibly survive the coming human decimation!

# B. We Have Met the Aliens, and They Are Us

Alien being cannot viably exist–cannot perpetuate or evolve–
because living being must have a lifeworld from which it evolved
in eco-relationships with other forms of life. For, a living being to
sever its relationship with the evolved system of life that created
and sustains it is to sever its own life and being. Any aliens from
outer space would perish on the long journey to Earth or, more
likely, even before the journey began. Moreover, to propose that
alien beings are visiting us is to propose that they are stupid enough

to leave their own lifeworld that supports them; or that they are desperate after having depleted their lifeworld and need exist in an alien, parasitic life attached to Earth-life, depleting and unbalancing Earth-life. But, the latter is a contradiction: for, if they are detached from their original lifeworld, they are more detached from and unviable within ours, and would seek out a lifeless planet to support their alien being that is alien to all life. Any aliens from outer space would know they are stranded outside all living being (or at least know the cruel unreality of naked life) and are doomed to cling to life without being. According to our knowledge of evolved life, this cannot occur evolutionarily or sustainably. They would envy Earthlifeworld and its world of many species beings; for, they would be creatures alive only–without any species being belonging to a home with other living species. They would be homeless life, life without any being and a field of being: such a life form contradicts the nature of life, especially mammalian, since life evolves within a world of ecological beings sustained within an ecosystem. Such alien life cannot occur evolutionarily or sustainably, because life evolves within a world of ecological beings sustained within an ecosystem. And yet we embrace this impossibility in our move out of Earthlifeworld into the alien being of sci-technic, commercial selflifeworld. Life must have being, being must have a world of diverse interrelational beings, from which to emerge and with which to sustain and evolve. Our modern, futuristic schemes ultimately entailed a 'second nature' that denies the immutable, evolved living nature of sentient, conscious life. Life must have being, being must have a world of diverse, interrelational beings from which to emerge, and with which to sustain and evolve.

## C. Aliens From Where, and How?

Alienism–alien phenomena and behavior attaining to significant, alien being within human being (alienized human being)–can be proposed as resulting either from a penetration and invasion from outside living nature (evolutionary life and its biosphere), that is to say, from outer space or the Earth's inner core. Or, as human authenticism proposes, herein, as having been created by unprecedented events within living nature's realm— from within human being (as alienized human being): living nature has experienced an event alien to her four billion year-long evolution.

The departure and transgression of the ego from whole self (and the dislocations, surrogations, rearrangements and dysfunction involved) depicts a key alien (phenomenon) as the origin of alienism. It is compatible with a critical mass-explosion scenario wherein anomalies incrementally increase to a point that excedes living nature's (evolutionary life's) capacity to endure or accommodate them, whereupon alienism emerges as unprecedented phenomena inside living nature, as an alien pathogenesis (analogous to malignancy, virus or infection) beyond her defenses or immune system. This scenario opposes the critical mass-impact account which has anomalies penetrating from outside (geographically or spacially) from outer space or from beneath the crust of the Earth. The philosophy, herein, proposes *increment-critical mass explosion alienism;* wherin anomalous pressure from unprecedented anomalies create a key alien that opens the way for incrementally emerging alienism to grow throughout the human realm to reach a critical mass explosion and transmutation of human being. Admittedly, this is quite close to saying that our human species has experienced an unprecedented, strange psychosis or that 'civilized' humanity is a monster to living nature and her evolution of life. And, since 'civilized' humans became the majority at some time during the latter half of the twentieth century, the 'madness' cannot be confined and restrained as psychiatric individuals are. There are two more distinctions to be made here. If the majority of a species is crazy–spiritual-ontologically pathogenic–this must be an unprecedented, *alien* 'madness'; and it follows that we lack natural defenses against it–still more evidence that it is alienism. Thus, our need for the deepest of human revelation and conversion.

It is beneficial toward understanding the existence and nature of alienism to momentarily see it as a malignancy of our species.

The emergence of cancer in people is associated with the aging process. Our human species is 300,000 years old. The increasing presense of this human pathogenesis, the alienism of 'civilization', across the Earth is in line with the common, modern idea (and prophecies) that a human apocalypse is near. The proposition in this book is that humanity is in the late autumn of its time and role in the play of evolution. But, the idea of a species malignancy must be scrutinized to determine whether it lies within the idea and process of evolution, or instead, is set against (is outside) the living nature of the evolution of life. No doubt, there have been some species that were able to recover from near extinction through genetic, mutational adaptation. But, has any species ever suffered from a pathogenesis that debilitates itself to the condition where it has the power and the will to destroy most of the life's evolved species and the ecosystems they evolved with and consist with! Can we really imagine that science (or philosophy), which is a major part of this alien pathogenesis, will seek to find and use a nonexistent vaccine to inoculate itself, its enterprise and 'civilization' against the alienism it disguisedly and inherently spreads throughout all Earthlife starting within humanity!

This book issues this challenge: let all those who affirm the legitimacy and reality of a 'second nature' and second human nature, who affirm and that everything in our world is natural and who deny alienism and the alien being within us; go these forth on a pilgrimage. With our precious television cameras hovering on their spirit, let them speak the doctrine of 'second nature,' or speak that everything including 'civilization' is a part of living nature: speak it to the clear cut mountains left nearly naked of life, to the dead and dying lakes and streams. Speak it to much of the Earth's species here before the Industrial Age, now gone forever. Speak it to the circus animals that the bars surrounding them are a line of trees to be scampered through to real life and freedom. Tell a million cows that the machines on their utters are calves ready to be nurtured by mother. Tell the amputees that their wooden and metal limbs are natural, can feel a loving touch. Tell the wheel chair people they are natural, bipedal human apes and can walk and dance. Speak it to the babies bottle fed that the bottle is the mother's breast. Tell the panic stricken infant in the store lost from mother that all the women are his mothers. Speak it to the children in the foster homes and in the jails, to the people in prisons, mental wards, in shackled down isolation that all is natural living nature. Speak it to the lonely souls in the city lacking loved ones to share a soul that must be shared to be human. After such a pilgrimage, no doubt, fewer could still speak the big lie about our human being; fewer could make the false leap in faith, this faith and belief in unliving, artificial nature and dehumanized, alienized being.

The proposition of alienism, it is clear, draws strength from a deeper look at evolutionary life. The life-force of evolution is inherently opportunistic and spontaneous; while the being-force is deliberate and exclusive. In family-community, social mammals and primates, family-community gene perpetuation takes precedence over the organism's gene perpetuation–reproduction–simply because the particular, species being is fulfilled *jointly* by *genes* and *nurtureculture* that are bequeathed in the family-community nature (and component) of the species. Some individuals are denied the 'right' to reproduce due to their insufficient ability to participate and fulfill family-community being. Evolution becomes that of the species' family-community paradigm perpetually fitting into its ecosystem of many living beings and the ecoregion.

'Civilized' humanity departs from perhaps fifty million years of its evolved, social, primate nature and five million years of evolved human nature: it overthrows family-community-ecosystem evolution of species being by subordinating organic, family-community reproduction rights and other rights of our species human being to (a) the egoself (individual) and (b) the oversized, anomalous, artificialized, social (and political) units of 'civilization'. Both the individualism (of egoism) and the socialisms (of collectivism) manifested by history (by 'civilizations') are departures from human and mammalian, natural evolution of family-community-ecosystem evolution. 'Civilization's departure represents the separation of life-force from being-force, their conflict, and the subjugation, debilitation and alienization of human being by alienated life-force. Our departure is not just from human evolution and human nature, but from social, mammalian evolution and social, mammalian nature. If evolution can and has thrown back 50-100 million years, this would render 'civilized' humans freaks of living nature. But, this is not precisely the case since no mammal or reptile has ever turned on living nature herself—evolving life—in the manner of systematic and accelerative murder of her species and ecosystems. The conclusion beyond an intelligent doubt (reason defends and rationalizes 'civilization') is that our departure, then, is from living nature's four billion years of evolution. We are emergent aliens to Earth's Earthlifeworld and its evolution of life. We must turn away from 'civilization' back toward home—toward our natural, authentic selflifeworld within Earthlifeworld.

We are not true freaks of living nature, but rather unprecedented, *alien* freak-monsters: unprecedented because we are cultural freak-monsters rather than genetic, created by the pathogenesis of our being through artificials—unprecedented, alien objects and phenomena—that are utterly outside of and alien to living nature's organismicity (living nature's ecological systems), to her biosphere and ecosphere. Analogously put, enough artificial chemicals and drugs going into our natural, human mind-body makes us physically and mentally ill: it is the same with other, artificial, alien elements-phenomena penetrating our natural, human species. But, we are beyond natural illness; our illness is unprecedented and alien to 4 billion years of evolutionary life. The exponential explosion of—or, at the least, the acceleration of—artificial, alien objects and phenomena experienced by humans in the modern age and climaxing in the Twentieth Century's ecological crisis is the creation of alien being within our natural human being. The proof is in the pudding; it is in our lifeway and culture: we do not fit into living nature's evolved system of ecosystems and as individuals we are falling out of family-community-ecosystem evolution (growth-fulfillment-perpetuation). Our growing, alien being within us is a growing pathogenic illness to all evolved, living beings including us. Alien being's destruction of natural being will be eventually reversed because four billion years of Earth's lifebeing-force is stronger and smarter than the alien-being-force that has emerged where it does not belong. The only question concerns the details, duration and extent of the coming human apocalypse. Because alienism has evolved in cities that characteristically destroy Earth's ecosystems of life, it seeks not Earth's living nature, her biosphere, her wilderness, nor human growth-fulfillment-perpetuation [eco-family-community evolution (sustainability or perpetuation)]. To the contrary, its drive is for artificial constituents, and, of late, even alien planets for the growth and spread of its alien being.

Evolution is only a one-way passage over the long passage. It can and has gone back to pick up belongings once tossed aside that are again necessary to the journey. 'Civilization', however, (the 'civilized' part of our selflifeworld) is outside living nature's evolution; we cannot, then, hope for a natural, atavistic recovery. Our recovery is found in the recovery of natural being, nurtureculture, nurturome, and selflifeworld (which is paradoxically unnatural since their loss is unnatural and alien); we must cleanse and purify these as much as we can of the alien adulterations of 'civilization'.

## D. Alienism Overrides Alienation

*Alienation*, plays a major part in critical psychology and philosophy, but misses the point and barely touches the reality of what is happening. It focuses on the subject's alienation from itself, another subject or social unit; the most particular focus is on the alienated's loss of feeling and identification. Even at its best, alienation, as we have conceived it, focuses on the alienated (troubled) relationships of these. But, *alienation* fails to come to grips with the question of what are the alienated relationships themselves alienated from. It is from this failure that alienation from self, others and social units remains essentially misunderstood. We fail to imagine and then apprehend that the alienated objects, subjects and relationships are each and all themselves alienized substantially out of their nature; and we fail to recognize a third alienation. No amount of therapy can begin to heal the alienated subjects and their alienated relationships without apprehending the third, wider and deeper alienation; such therapy can only apply bandages. We are required by our age of critical juncture to recognize the deepest alienation, i.e., that of urban, 'civilized' selflifeworld away from and out of its natural, real, authentic, archetypal paradigm, out of our species Paleolithic selflifeworld–out of our human nurturome–evolutionarily intergrown within Earthlifeworld; even though this realization delegitimizes, on the deep level of human welfare and sustainability, modern (and urban based) therapy along with 'civilization' selflifeworld itself, which ensnares us. We are required, at our critical juncture in human being's journey through time, to recognize this deepest alienation and, through recoveries of our human beingness, experience, again, renurturalization and healing.

# CHAPTER VIII

# HUMAN AUTHENTICISM VS. ALIENISM

## A. Short Forward: Reaching for Stars

IT IS HUMAN to be awed by the firmament, to even reach for the stars through the imagination (as in astrology). But in our age, this natural, celestial mysticism has been replaced by some with the concrete goal of pioneering and colonizing the alien planets of outer space. This goal is new and alien—fundamentally, inherently outside of human consciousness, mind and spirit. Nevertheless, one sci-technic enterprise, the "space program," (and allied enterprises), proposes that we incorporate it into the human spirit and agenda. In truth, it is an alien dream indigestible to natural, authentic, evolved nurtureculture, nurturome, and genome—human being, reality, and purpose—and self-destructive as a goal.

After 300,000 years of our human species' journey upon the Earth, the wondrous jewel of the reachable universe: who are these amongst us—and boring inside us, into our human being—that propose the folly of an unworkable, alien, human transmutation into alien beings, permanently colonizing Mars as their home? Who are these that propose we pioneer terrain alien to the Earthlifeworld—biospheric or ecospheric life—that created us! And alien to the world (web) of life and field of being we were evolved, ecologically, to fit into by Mother Nature! Are they a new species, these that reject both human being and human, evolutionary creation in favor of alien being and the alien *re-creation* involved. No, not a species this, but merely an infected portion of our species. No human species can emerge from sci-technic and commercial creativity.

This is emergence of alien being out of alien mind, spirit, being, sci-technics, commercialism and power-wealth ethos; it grows like a tumor within the selflifeworld of our, natural human being. This emerges much as a cancer or virus emerges in our body and attempts to take over and destroy human cells, organs, life and being. Just as biologic cancer must be turned back, this drive into alien being by some alienized minds and spirits (captured by ambition alien to us) must be turned back. (For a thorough refutation of the space program, *see* Part 6 of Ch. X.)

Shall we, then, scuttle this alien dream! Shall we embrace our living human being, and spirit, nature, and reality, and protect the natural sources of it–natural, whole self, natural family, natural community and the Earth that created these! Or are we leaving and forsaking our real, primordial mother, Mother Earth to live with the unreachable cosmos of Saganism on dead planets, while we suck the last life out of our living Earth? Have we forsaken the real nurturing, growth, fulfillment, and perpetuation living nature has provided since slowly bearing us forth from evolution into her four billion year-long creation-play of life!

## B. The Main Conflict and Struggle

In our time, our human being and spirit are battle fatigued from continual evasions and rationalizations of our natural, authentic, human truth and reality. We have come to where we either conditionally surrender to the alien forces, objects and phenomena we have incrementally created, which have marked the infected human purpose of modernism's Homo alienus; or we receive–through our remaining natural, authentic intelligence and

consciousness–the human revelation of what we are as human being and our rightful destiny from ancient, primordial legacies; and,
then, as much as we can, regenerate––renurturalize and renaturalize–
natural, real humanity with its natural, human selflifeworld and
Earthlifeworld. We have come to this last chance epoch of our
humanness, surely, to meet head on a new force preying within
our primordial, 300,000 year old humanity. But, how many of us
will set ourselves down to the cultivation–solemn, desperate but
magnificent–of The Last Human Spring?

Our creative evasions of our true and real humanness have put
crowns upon imagination and curiosity, pressing them into transmuted, quasi forms. We have created and overvalued new sacra-
ments for our new quasi religion–for science-technocism and power-
wealth-spirit-ethos. Our human intelligence, consciousness and
purpose, once subjugated by royal elites, monarchies, or facsimilies
of these, now are driven ever deeper toward a thorough subjugation and enslavement by the sci-technic commercialism of the
power-wealth ethos and dynamic. Lately, we have set about creating and welcoming into our alien, quasi self, life and world personal computers and computer-robots and almost any creation
that shows how powerful sci-technics and commercialism is, the
highest power visibly, concretely and practically, manifesting itself
in the human realm. It has become our practical, concrete deity,
and, according to our false perception, demonstrates human power.

It is evident that we are approaching the end of our divergent
road away from living nature and human nature into unhuman,
dentured, and alien being. We even prepare to migrate our bio-
logic-intellect-egoself to an alien world more alien than cities, i.e.,
the moon and Mars, attempting to place that half human, half
alien self into a wholly alien lifeworld–into an unlife/unworld.
Attempting, as the apostle of sociocultural transmutation and decimation, Carl Sagan, put it, "to extend the human presence into
the solar system" (*See* "*The Cosmos* of Carl Sagan Vs. Earthlifeworld,"
Part 6, Ch. X.), not knowing or denying through existentialism,
that human being has a nature which entails and requires living

nature–*earthly* selflifeworld. (Ironically, Sagan himself in the last years before his death in 1996 softened and balanced his alienism with an increased appreciation for the Earth as our magnificent home. Ironically, Sagan departed from the ranks of true Saganites.) Saganites are half alien already; and, any Earthlings outposted to colonize an alien world into some quasi 'lifeworld' will be three-quarters alien, with only their biological portions remaining generally human. For, in the same way that the American colonies' allegiance soon shifted from European lifeworld to American lifeworld, the alienized, quasi, intellect-egoself, according to Saganism's adoption of infinite, human transmutation (possibilities), will soon identify with alien, quasi unlife-unworld, thereby forming human body 'grafted' to alien, intellect-egoself unlife-unworld. Saganism is markedly alienism and seeks to create through an adventure/discover spirit-ethos and social transmutation, its version of the alien, space creatures that alienized Saganites long to find and mingle with. Having lost or subjugated their allegiance and identity to their ancestral humanity, their allegiance and identity go to postcestral alienity–to being after it has mutated away from and generally out of human being, which they falsely believe to be feasible and heroic. They are emergent, Homo alienus traitors, not of their country, but of their human species and human family–of (over a million years of) human evolution, and traitors to four billion years of Earth-life!

The emphatic recognition and understanding of alienism is crucial both to this volume, and to the preservation of humanity. *Alienism* is both the growing, artificialization infecting humanity, and the new mega-philosophy formed through the coalescence of all the modern, contemporary ideologies, sciences, and philosophies. It is the condition or state of alien being that human being is unknowingly moving into via pathogenesis with alienism. There are two ways to move or flee an unsatisfactory present state: into the past or into the future. Humankind can either preserve Earthlifeworld and effect some recovery of its natural, authentic, human selflifeworld, its structure-function, systems and human elements-phenomena that constitute the nature, identity, being, spirit and reality of its 50,000 year old modern humanity; or it can continue the movement to incrementally abandon its human species' selflifeworld, attempting to create what human authenticism reveals in this volume to be impossible–alien being as steward of Earth; and, therein create an alien egoself, life and world to displace natural, human selflifeworld. Human authenticism is our being-force and its philosophical movement to gradually rediscover, recover, resecure and preserve our natural, authentic, modern humanity and its indispensible nurtureculture; alienism is the movement to incrementally abandon them in favor of emergent alien being and culture.

Of course, the dialectical struggle–of natural, authentic human nature, being, spirit, identity, reality, human selflifeworld and Earthlifeworld versus alienism's adulteration, artificialization, and alienization–began occurring as skirmishes with the emergence of 'civilization'. Now the revelation of human authenticism amidst the accelerated growth of twentieth century, sci-technic commercialism is tantamount to revealing a foreboding human decimation. And, in a real sense, it is a war, our human species' social-spiritual-ontological war against the advancing malignancy or pathogenesis of alienism. The consolation for the colossal, ugly struggle that has now entered its decisive epoch is the shining vision we now behold more clearly through the human revelation: magnificent, natural, authentic humanity's nurtureculture and selflifeworld. Further, there is the consolation of knowing that when Earthlifeworld and human selflifeworld are secure in a state of preservation, it will have been done through what remains and can be recovered of our natural, authentic humanness and human being. If this Last Human Spring fails, we fail millenia, even millions of years of ourselves–our ancestors–and our duty to perpetuate our legitimate, time-proven humanness.

So, readers, choose your sides, now that the human civil war within us and our lifeworld is revealed; and choose with this understanding: unrevealed and unrecognized, the tide of this 5,000

year old conflict has been, with some defeats and setbacks, with
fallen, 'civilization' humanity, as the force of 'civilization'–alienism
and its quasi synthesis–has been gradually overcoming the lifebeing-
force of human nature and being, effecting cyclical retreat and
destruction of living nature, human nature and being as the artificial divides and conquers a portion of each natural thesis and, likewise, each succeeding quasi synthesis. But, whereas 'civilization'
organizes synthetic humanity and disorganizes natural, authentic
humanity; once this is realized through the human revelation, the
struggle is recognized as a primary, underlying war of real, natural
human being against the alienism of 'civilization'. And this opens
our consciousness for a declaration of spiritual-ontological war in
the minds and hearts of many of us. It inspires organized defense
and resistance, and brings recovered, natural, authentic human
being, spirit and purpose. Conflict attained to spiritual-ontological, cold, civil war brings both the strongest and the highest purposes of human nature and being into this last human struggle–
the will to live and the will to be (to be human being) respectively.
Both sides gain human purpose–synthetic humanity being partly
human gains part of it; but, the renaturalizing and renurturalizing
of The Last Human Spring summons all of human purpose and
spirit in this conflict attained, through the human revelation, in
our age of looming, manifold decimation, to a division within the
human soul. It comes to this, then: the force of 'civilization', of
history, of alienism, and its alienized, infected human being against
whole, human nature and being, against a human nature and being now with the will to survive and the will toward being unified;
as living nature grants us our Last Human Springtime. It is organized, natural, human will to live for human being and soul versus
organized, alien, intellect-egoself will to live for power-wealth. It
is human will to live *and experience* the growth-fulfillment-perpetuation of human being versus unhuman, alien will to create
power-wealth and dominate natural human being and living nature. Since natural, authentic being only wills to be, fulfill, maintain, perpetuate and reproduce its being, it is evolution's human

being versus digressive, power-wealth, alien being. It is the natural versus the alien; nature versus unliving nature. It is human, *life-being-force* versus unhuman, *life-power-force*.

The question remains: if synthetic, quasi, alienized being is diminished and adulterated into partial human nature and being, is the portion of human being thereby lost, (a) replaced with sub or pre-human, atavistic being, (b) replaced with an improved be-ing, (c) replaced with equal but different being or (d) merely lost and thereby leaving less human being. To those well within 'civili-zation'–to blended, 'civilization' humanity–and most reading this are significantly to largely so blended–the answer is partially (a) and generally (d). For, there is nothing to be gained in the way of 'human' being upon its adulteration with a debilitating patho-genesis or malignanacy. The mind trained into alien, artificial thoughts and behavior gains adaptation to alienized, artificialized, quasi 'culture', but loses both some natural mind and being that fit their displaced, natural, authentic nurtureculture within Earthlifeworld.

Some readers, surely, will protest. Afterall, a key feature of post industrial society is the rise of the term and concept, *super*, as in superman, superstar, superbowl, super anything that is regarded recently superior. And, further, sci-technic humanity is especially opposed to believing it is inferior or is in decline. Indeed, how is human authenticism to persuade the 'super people' of 'super in-dustrial' or high-tech society that they are not 'super people'? Most cannot be *persuaded*; for, one either 'knows' from some retained, natural intelligence that something is missing within us and be-comes a seeker, or one does not have this retention of natural intel-ligence. Those that do have the basis to receive the human revela-tion of the real, the good and the pathway to it.

The more precise answer from the philosophy of human authenticism regarding the above question of the fate of future human being, is that both species life and being must evolve or maintain its ecological fit, or it will throw back or move toward extinction. Because synthetic humanity is partly natural, partly artificial, it is both throwing back to Homo erectus and Paleolithic human being, acquiring atavisms and decaying from alien pathogenesis in the process. (The present population of six billion people includes both alienized—civilized—humanity and a'civilization' fork humanity. The former is headed toward a decimation that may include a rather sudden apocholyptic event, but will in itself be slow, lasting perhaps two to ten generations. Since a substantial portion of overpopulated, natural humanity are dependent upon artificial support of 'civilized' humanity, the decimation will include both groups of humanity.) That is to say, synthetic humanity, being part artificial–nonnatural–cannot follow living nature's evolution for natural species. The natural part of it is throwing back because of dilution and adulteration breakdown of species nature and being; while the artificial part of it is dying of alienism, for the artificial is non-natural, non-participative, non-functional in natural human being and spirit; it is decayous or cancerous to natural, human being, spirit, reality and selflfeworld. Increasing progressive synthesis is progressive decay. Complete synthesis is complete death to people and extinction to humanity. The 'civilization'ization of humanity, is its incremental removal from living nature and from human nature and being through humanity's adoption of unnatural, unauthentic, unhuman, alienized culture. This synthesization of humankind, in reality, is the manifoldly disguised decay of our species.

Declining with the decay or cancer of artificialization and alienization, humanity senses or feels danger, is trapped in the age of anxiety, and retreats ever deeper into biological intellect-egoself and its constructed quasi life and quasi world. We survive but lose some of our human nature, identity, being and reality. The decay continues with succeeding generations born into our species', degenerative social and cultural disease, which is hidden by the so-called 'growth' and 'progress' of urbanization, diverse organizations, material wealth and feverishly cultivated, escapist entertainment. This 'growth' and 'progress' are, in reality, the alienizing, cancerous, elements-phenomena and constructs infecting natural, authentic self, family and community, intergrown with Earthlifeworld as our selflifeworld. Humankind trapped in incremental, progressing decay creates psychology, social science and philosophy to understand, rationalize, accept and treat decaying humankind. We of 'civilization' make a life of understanding, caring for, treating the casualties, winning battles, making compromises, enjoying quasi tribulations, and celebrating the decaying human condition we so creatively decorate and masquerade as artistic and sci-technic human progress or cultural evolution.

# CHAPTER IX

# WALDEN III: AN ECOPHILOSOPHY

## A. The Exodus From Earth-Being

THE EARTH'S ENVIRONMENTAL crisis is even more serious than deep environmentalists have understood it to be. The evolution of life has been disrupted and reversed by modern 'civilization'. Human being has been diminished and debilitated. Life's abundance and rich variety, and its diversity within each species is being ravaged and homogenized by sci-technic commercialism, by the scheme and dynamics of alien 'civilization'. The shit of history is hitting the fan of evolved Earth-life.

More precisely, alien being is growing within human being: the explosion of artificial objects, phenomena and behavior alien to natural human being is transmuting human being into alien being. We are rapidly becoming aliens to Earth-life. We both deny and display our alien being in our fantasies of, or belief in, aliens more alien than ourselves. (*See* CH. VII, THE BIG BANG WITHIN EVOLVING LIFE: THE ORIGIN AND EMERGENCE OF ALIENISM.)

We misunderstand our ecocrisis with our belief that it is our actions of pollution, our unenlightened habits, and ways of thinking that must be reformed. While true, this belief misses the systemic nature and depth of the ecocrisis. *The destruction of the natural world eminates from what we are as alienized human beings–from the kind of beings we have become. The environmental destruction is primarily within the environs of our human being–the destruction of living nature's natural human being, her human nurtureculture, nurturome, and selflifeworld precedes and effects destruction of the natural world enveloping us. When we come to realize that the pollution of human being and soul–material, biological, social, cultural, spiritual–ontological–is primary to, and the source of, environmental pollution, we will have broken through our illusion, and can begin the real fight to preserve human nature, all living nature, and their world– Earthlifeworld (ecosphere).*

## B. The Innate, Ecological Paradigm

Earth-life's ecological paradigm can be and is supported by re-cent science; but the base and source of it must be understood to be our reawakening of natural consciousness of natural human being's participation in and dependence upon living nature within us and enveloping us. Our grasping is a grasping by our very Soul for the reexperience of this consciousness–not a grasping for newly discov-ered facts that scientifically or philosophically validate the real, an-cient grasping itself. This reawareness, this recovered consciousness, has occurred episodically in cultures and perennially in individuals throughout history, usually, in spite of and without science, philoso-phy, and reason. It is a deep grasp impelled by instinct, intuition, and the unconscious, into ourselves, into humanity, into our spirit and being. It is only secondarily a grasp of intellect and reason into science, philosophy, and theology. The ego and egoism of the intel-lect-egoself, along with the power-wealth ethos, largely obstructs any reachings of reason, science, philosophy, religion, and their enter-prises back down into our natural being, spirit and mind. Our natu-ral, whole intelligence, self, will, and purpose of our species' human being inherently precedes, delegitimizes, and opposes the intellect-egoself, its intentions and enterprises.

Modern science merely goes along in its vehicle for its ride into rediscovered terrain of natural human being; it must not be credited as the vehicle, nor substituted as a new base. For, the grasping back into natural consciousness of whole self and being has been occurring since 'civilization' began pulling us out of these naturals that living nature, through genes and nurtureculture, our nurturome, bequeaths us. *It is an act of human nature to seek natural human being and its consciousness. Instruction is secondary to example. It is not instruction, but renurturing and natural, spontaneous behavior that reawakens the soul's hypnotic, intoxicated, addictive slumber induced by 'civilization'. In a real sense the ecological movement is a reaching of natural human spirit and being up into science and 'civilization' toward unblocking the accumulated obstructions bequeathed by 'civilization' to individuals, social units, and cultures.*

## C. Ecological Identification Without a Base

For ecophilosophy, the way to reconcile the competing claims of striving individuals is to recognize that the 'individual' of individualism, in and by himself does not experience nor manifest *all* of human being nor fully represent human being. Instead, the triecologic unity–organismicity of ego-self, family (-self) and community (-self) involved in a natural lifeworld within living nature–is natural, authentic, human being; this whole, human being precedes the organism (individual being) of human being. Human being can do without any particular individual, but the individual (organism) cannot do without the rest of his/her species' human being, intergrowing, fulfilling, and perpetuating the organism's portion. One cannot learn and play out language and the social relation of human being by oneself. In natural human being and community there are only incidental, competing, conflicting claims amidst individuals. The 'individual' is in harmony as part of human being, spirit, identity and reality. Competition remains naturally resolvable instead of the problematic kind it becomes when 'civilized' society transforms it into a social principle that then guarantees classes, factions and conflicts. In short, the social, spiritual-ontological triecosystems of bioself-family-community–growing-fulfilling-perpetuating itself–in a natural ecoregion is the true, authentic structure-function-dwelling of human being rather than the individual organism or the overpopulated city or state.

True, ecological identification has no particular, paramount base. We don't have to find a personal base, social base or natural, non-human base within natural human consciousness. Neither the personal self nor the self-appointed, scientific community-self is the natural, human self and its consciousness. There is no *base*: that is the point. Natural human (and social primate) being has no base. Would not such a base deny the organismicity paradigm in favor of a possibly mechanical structuralist functionalism! It is only from the mind-body organism as the reception center of the senses, thought, feeling, and instinct that we, as intellectual-egoselves (entities of reason), look for a base or main site for species human being. It is especially our modern illusion that the individual is the origin and base of these; for, without family, community and lifeworld, senses, thought, feeling, and instinct could not unfold. In our human case the reception center is not the base or *basis*; since the organism, family, community and much of evolved life's Earthlifeworld, human ecoregions–are all indispensable, equally vital, basic (base) parts of human being. *All parts of human being are created equal, and endowed by evolutionary creation with the inalienable right to play their evolved parts in human being. Thus, only human being so constituted can play its evolved part in Earthlife's world (field) of evolved life and being.*

Something *very unnatural* to our human species has recently happened to effect the perception and practice of basing human being in the mind-body and intellect-egoself (and generally limiting it there). In living nature we have to go as far back as non-social (non-family-community) mammals and reptiles in order to find such animal being. Does the intellect-egoself (and egoism) of the city, then, represent a long evolutionary throwback? Evolutionary throwbacks usually are to the previous species, to the genus or family. If we are witnessing an evolutionary throwback, a natural phenomenon of living nature, it would be rare to say the least, and, far more likely, unprecedented and alien. It is a much more credible and intelligent conclusion, upon considering the degree and exponential increase of artificial, alien elements-phenomena represented by the artificial alienism—of the egoself and its quasi life and world, so progressively destructive to whole, human self, whole human being and living nature herself–that this is ultimate alienism, i.e., alien being that is outside both evolved Earthlife and her natural human being.

There are, of course, many transmuted, sci-technic people (emergent Homo alienus) that welcome the idea of moving outside or *beyond* human nature and natural, human being, whether or not it is also a movement outside living nature herself. However, since this enterprise of alienism—of increasingly alienizing human being–is outside the natural ecosystems of our four billion year old, evolved Earthlife, and outside our human species' 300,000 year-old role in Earth's ecosystems, the chances for alienism's success are nil compared to our chances for human fulfillment and perpetuation upon staying generally within our beautiful, successful selflifeworld within the Earthlifeworld of Earth.

## D. Neither Ecocentrism nor Anthropocentricism

Our ecological quest for natural human being need not center living nature at the cost of the *natural* human realm. The human realm in full reality–before 'civilization'–is our *natural and human* center: it is within and fully interfunctional with–in no opposition to–living nature. We as a species (as opposed to 'civilized' humanity) are nature within nature. We are all natural, enveloped, cradled and grown by living nature from both the 'outside' and 'inside' of the human realm because we are a *small* part of nature, within life's larger, more eternal reality. (The primary, revolutionary, human revelation that our species is inherently, wholly natural is presented below and in Ch. II and III.) It is our species

self-consciousness of and identification of/with its particular, species being, and the biological component of our human being that
make us our most *immediate* concern. It is only from this imme-
diacy that *anthropocentricism* gets its episodic and limited legitimacy. The dictionary's definition is misleading because the human realm–before the pathogenesis of alienism carried by 'civilization'–is a part of nature within living nature. Those who would
preserve the natural world and humanity must realize this and all
its ramifications through a deep, ultimate, human revelation.

*When* the *relationship* between the two realms, i.e., (a) living
nature's all natural human realm within (b) living nature's realm,
is seriously debilitating both, then both and their relationship are
the concern. However, since it is we that assault and violate
Earthlife's health–her ecological, organismic world (field) of being–through the alien pathogenesis emergent within us as the alienism of 'civilization', we must diagnose and treat the illness that
causes our departure from our home within living nature. Our
sacred trust–given from evolved Earth-life and her human being–
is to heal what we can of our infected human being, nurtureculture,
nurturome, and selflifeworld, to work toward healing the alien
illness transmitted and bequethed to us by 'civilization'.

*So, what is the ultimate, inherent relationship between the di-
lemma, 'civilization', and a four billion year-old living nature that
includes a 300,000 year-old human nature of Homo sapiens? The
crux of any, significant, human redemption hinges upon this apprehension, this deep revelation. We must know what living nature really is,
what human being really is, and what 'civilization' really is, in order to
know the real relationship of these.* A small minority of us cling to
the belief that 'civilization' is somehow a new part of evolved human being within living nature. Most, however, accept modernism's
belief that 'civilization' is cultural evolution upward, an improvement beyond and above pre'civilization', its prehistoric culture,
and lifeway, or simply a mysterious, cosmic manifest destiny for
better or worse. The last two beliefs reflect our loss of our *will to be*
our human being, as our will to be is gradually displaced by the

will to recreate our being toward a presumed, existentialistic, sci-technic utopia or a misconcieved cosmic destiny. *But, the final lesson learned from the dilemma of 'civilization' is that our only real, authentic fulfillment and perpeptuation as a species lies in Earthlifeworld, not in the universe-cosmos; that our redemption lies over our shoulder, behind us, in our 300,000 year-long, evolved role in the play of evolutionary life. We will finally learn that the will to live (life-force) cannot be separated from and given domination over the will to be (being-force). We will learn that, ultimately, we cannot sustainably embrace anything else but the human being living nature has fashioned for us in her very slow changing play. For, we cannot perpetuate our human life and continue the human journey, without our human being staying sufficiently human so as to stay in the only, real, long-running play humanity has a part in. No man is an island unto himself, and no species a play or sustainable world unto itself.* Living nature strikes down or out any species destroying the other players and the lifeworld that the rebel species unknowingly needs to perpetuate its own life. Contemporary, 'civilized' humanity's conspicious self-destructivity comes not merely from having lost natural, human intelligence, but from having alien intellect in the role of human intelligence bonded to the alien egoself that thinks it can continue to alienize its diminishing human being out of living nature's play into a new play of an artificial, second human nature within a 'second nature' of artificial world. We must pull back from our assault and violation of living nature, human nature and her other, natural, living beings or she will eventually deci-mate us. Living nature's treatment of the human illness within her is her great compromise with alienized, 'civilized' humanity—and our basic, ecological remedy—is Earth-folk culture and lifeway. We must embrace her 5,000 year old self-treatment.

These three beliefs, often intertwined: that 'civilization' is (a) somehow a new part of evolved human being within living nature, (b) something *culturally* evolved through a 'second nature' and human nature that is superior to our first original living nature, or (c) a cosmic destiny beyond living nature; these beliefs fly in the face of the fact that everywhere on Earth 'civilization' has emerged, it has been destructive to living nature's ecosystems of life and to human family-community fulfillment-perpetuation-harmony within living nature's systems of life. It is at this step of our eco- logical quest that we have blocked a crucial question: *why did liv- ing nature's evolving life, after four billion years, and after three mil- lion years of ape-human evolution, create 5,000 to 10,000 years ago (in one minute of human time) a human socioculture, mind and being that is destructive to living nature, its evolved ecosystems of life and its natural, human being?* Some thinkers in the deep, green move- ment have asked this question in more simple, less suggestive forms: Neil Evernden laments, about the ecological crisis, that nobody asks "the question of ends–why we are doing all this in the first place"?[1] Max Oelschleager mentions "the philosophical questions posed by Leopold, Evernden, and many others. Namely, why does Lord Man do what he does in the first place?" Oelschleager con- tinues, "by raising such questions–*even if unable to answer entirely the formidable, philosophical problems* raised by its perspective claims– ecocentricism directly confronts resourcism and goes beyond the cognitive bounds of preservationism."[2] Is the answer so radical that it is beyond even the deep or radical eco-awareness? Should the question go deeper–to the legitimacy and reality of 'civilization' itself? *Could it be that 'civilization' has emerged within life's evolved human nature as a contagious pathogenesis alien to her immune system (her evolutionary defenses) and alien to the lifebeing-force of Earth's evolving life! Is this not a cosmic event comparable in meaning and consequence to the creation of life and its evolution, an event that marks the reversal of evolving life back toward inanimate matter through increasing extinctions of species and the breaking down of eonic, eco- logical complexities of living nature's evolved life, its systems and process!* The destructiveness of 'civilization' upon life's evolution and 300,000 years of evolved human being is very strong evidence that 'civilization' is alien to the human species and family, and to Earthlifeworld itself. This book's philosophy of human authenticism reveals that 'civilization' is, indeed, alienism; that we, infected with a growing, contagious pathogenesis–alienism–are the real aliens upon the Earth. This philosophy, in addition to affirming a completely natural, Paleolithic culture in no conflict with natural evolution, proves beyond an intelligent doubt (reason is a defense and rationalization of 'civilization') that Earth's lifebeing-force has been penetrated, violated, and interrupted, effecting a schizm that separates life-force and being-force; the two of which cannot be viably separated in living nature's human selflifeworld within evolution's Earthlifeworld. (Delores La Chapelle, for one seems to point suspiciously at 'civilization' with her conclusion that we need to inquire more deeply about just what is the nature of the enemy of living nature, the ecodestruction of which we fight against.)[3]

# E. From Wildernessism to Ecophilosophy of Human Authenticism

## 1. Resisting the Exodus from Earth Village

We are greatly indebted to the splendid wilderness philosophers, deep ecologists, ecofeminists and others in the green movement that have struggled with significant success to divert 'civilized' humankind from an ecological decimation. But, they still cling to some obstacles that block the transformation or recreation of 'civilization' they pin their hope upon. They make it to Walden Pond–to wilderness consciousness, to Paleolithic, or Neolithic, Earth-goddess consciousness–and experience moments of partial, natural, ahistorical consciousness. Seasonal or weekend journeys to the pond (or the mountain)–the wilderness–maintain a link, a crucial identification with wild living nature. Visits to the wilderness even seem mystically to smack of animal migration and Paleolithic, seasonal journeys from camp to camp keeping to the cycles of game animals and gatherable plants. Can those who love the wilderness and Mother Earth, a small minority, save the wilderness, and therein, humanity from a manifold, unnatural decimation? Or, is 'civilization's movement, dynamic and purpose that very manifold destiny we, in our most sad and exasperated moments, fear it is: the effective decimation of living nature's wilderness and her natural human being?

*Have we not been camping upon the threshold of the deepest revelation of our natural human being? The revelation just before us is that we must return again to the acts of love that make love with the wilderness without and within us: make camp, make love, make natural family, make natural community, make our recovered home, and remake the wilderness commons of the living nature-human relationship. We must reconstruct it as both the wilderness lands and living nature's natural human being that has cultural boundaries and patterns that must be kept within (wild, pure) living nature and her (pure) human nature, and preserved as our wilderness commons of Earthlifeworld. We 'civilized' humans must recover the ability and passion to make love with living nature and to mate with living nature—mere biological reproduction without spiritual-ontological and spiritual reproduction does not perpetuate humankind in the journey of human being through Earth-place and time. This is not an incest–of biological sex–but the spiritual-ontological reproduction of natural, species being with the body (world or field) of Earth's living being. This is also deep, whole self-love, since natural human being lives and dwells within living nature's Earthlifeworld of being. Human being has a triecological self–biological, social and spiritual-ontological selfhood.*

But how can we recover our love relationship with living nature's Earth-life, which includes natural, whole self! After all, 'Civilization'–especially the Modern West–deliberately and systemically estranges us from Her *and our natural, whole self within Her! We certainly must do a recourtship of evolved, living nature–Earthlifeworld—which includes human nature evolved within her. We start with recourting our natural, whole, human self and being. But, neither our human part of living nature–natural human being, soul, and whole self–nor living nature herself, will respond to our courting overtures so long as we continue our earnest*

*relationship, our loving partnership with 'civilization'. Our spiri-*
*tual-ontological polygamy—our betrothal to both evolved life and*
*'civilization'—blocks the recovery of our loving, nurturing relation-*
*ship with living nature's Earthlifeworld.* Our two loves are aggressive and deleterious to each other, as evident by the debilitation, and considerable death, of much of living nature and
human nature, and by the declines and falls of 'civilizations'
effected by the defensive blows and retaliations of living nature
and Her human nature. Living nature will not and cannot accept us back if we bring Her violational enemy, 'civilizaton',
with us. *Any sustainable, harmonious return to living nature's*
*Earthlifeworld and to our natural, whole self and human being*
*must be made either without 'civilization' or with our relationship*
*with 'civilization' generally, systemically subjugated and secondary*
*to our relationship with She who has created us from 4 billion years*
*of ecological life, purpose, meaning, and being, and who still sus-*
*tains us, despite our abuse and destructiveness of Her and our whole,*
*human self within Her. We need a new version of a'civilization',*
*Earth-folk socioculture that moves into repurification—into less popu-*
*lation and adulterating artificialization.* We can't go home again,
but only move back into recovery and preservation of some,
natural human being and its home selflifeworld.

Are not the streams that flow to the sea the blood vessels both of
living nature's Earth-body and of natural human being? Are not the
sea and sky also the heart and lungs of our human being moving in its
time journey within Earth's world of living being? She has born us
forth with all her life from *Her* star's light piercing *Her* dust-flesh and
water-blood body. Her vital organs, flesh, blood and spirit are also
those of her children. Can we really continue this exodus out of this
Mother of the Village of Life and all living beings! Despite our deep
denial of what we are doing, to leave this ecological, village community of all living being is, in ultimate reality, to leave human being
and all living beings. But, we proceed in doing just that. Hence, the
human species has committed the taboo living nature sets against

every species. Hence our exodus out of Earth Village into alien being, self, life and world.

## 2. Paleolithic and Neolithic Lifeways vs. The City

If the emergence of agriculture was, as some ecological thought (notably that of Paul Sheppard[4]) holds, the original fall from living nature, from Paradise, or from Eden; nevertheless, it remains true, as Sheppard admits, that it is a long way (through time and in kind) from tribal village and family farming to the coercive, subjugative, organized agriculture of the first cities, nation-states, empires and modern corporate empires that fragment the organic family and community on the sacrificial, power-wealth alter of some king-god, royal class or technocracy. Can a deep ecologist or ecophilosopher, embracing Paleolithic or Neolithic consciousness—from the backdrop of the city's comprehensive artificialization and dehumanization (and out of the artificial university that generates that city)—visit a Hopi village or any other family-community centered, low-tech farming way of life, and pronounce such people "fallen people?" They can only do so upon a technicality at best. Such organic, family-community farming is ancient, highly natural "bioregionalism," and must be ranked quite high by any true measure of natural, authentic, whole human being. For, this resistant humanity has not experienced the *big fall* into city 'civilization'. The city lifeway (and consciousness) is ten to one hundred times more removed from Paleolithic humanity's unity within (wild, pure) living nature than that of family-community farming. To extol or idolize Paleolithic lifeway and consciousness at the expense of the Neolithic, which is substantially carried on and represented in Earth-folk cultures and lifeway, is to overlook and miss what opportunity we have left on an overcrowded Earth for a partial redemption for a fair part of humanity. This happens to deep, ecological thought that has not completely shed the fundamental value driving the idea of progress: novelty. This is to say that while embracing prehistoric or premodern conciousness, we are compelled, through the modern ethic of creativity, to fashion it compatible with overwhelming, modern 'civilization'. In this way we deny that we are overwhelmed and doomed by our addiction to modern 'civilization'. We forget that the first step toward recovery from self-destructive addiction is the admission of helplessness, and the need for help from a greater power than modern 'civilization'–help from a force that preceded all 'civilization'–the spirituality of our evolved lifebeing-force. In practice we must assume a position in compromise with the modernity that enfulfs us. But in principle, in full reality, there is no compromise: we must work to recover, cleanse, and reconstruct the elements of whole human self and being–to heal our spirit and lifebeing-force. Our partial redemption is a respringing from within our addicted, Human Soul.

We of ecological consciousness, of the Last Human Spring, can aspire to the Paleolithic spirit in our heart and soul; but, it is to keep alive that pure, human being and spirit buried within us, and serves to recapture moments of our natural, original spirit and being. In the bulk of our practice and lives, our appropriate, feasible aspiration is to practice this wild, pure human being as it has been substantially carried on in Earth-folk culture and lifeway, largely in harmony with living nature, and more accessible. The long leap back into Paleolithic spirit and consciousness is more achievable by first stepping back into the Neolithic spirit as substantially represented in authentic, Earth-folk culture, lifeway and consciousness. Humanity's substantially reachable aspiration is a revived Neolithic spirit within a recovered Earth-folk lifeway. Our partial redemption lies therein. We need a post-historic Neolithicism that aspires toward and partially recaptures the lost Paleolithic held sacred.

## 3. Post-Historic Primitivism Within Earth-Folk, A 'Civilization' Paradigm

To talk of substantially recovering Paleolithic consciousness and values (primarily through experiencing the wilderness area) is safe philosophical ground because the Paleolithic way of life is inaccessable; and, its admirers are often hailed as a new breed of romantic poets and philosophers. To help secure this sanction from the establishment, they give new rhetoric to their rediscovered old ways of prehistoric and counter-historic, Earth-folk living. Back to the land is dubbed *reinhabitation*; the family farm and organic community becomes an early form of *bioregionalism*. It is easy to sympathize, even embrace, this customization of old truths and realities with the dress of contemporary, ecological science and philosophy. These professionals, too, are lost and pushing home-ward in the particular way they feel they can. Their dilemma is largely ours. Tragically, such talk about a new, paradigmatic ecocentricism and/or a post-historic primitivism flies in the face of that which preceded and outflanks these contemporary, concep-tual innovations, i.e., a 5,000 year old, substantially successful, counter (socio)culture. It does so by failing to acknowledge, de-scribe and emulate this Earth-folk counterculture, its unorganized, partially unrealized, heroic struggle against 'civilization's' progres-sive, ecological destruction of the natural world and natural, or-ganic, human family-community. In a meaningful sense, folk, a 'civilization' socioculture can be viewed as prevailing over the ur-ban culture of modernism right up to the allied penetrations of the automobile and television. World War II sci-technics united with postwar commercialism made more potent by television, be-came the invasion force launched by modernity to more completely subjugate global, Earth-folk societies to 'civilization', and modernism's "progress." Prior to TV the spread of literacy was in-sufficiently complete to break humankind's clinging grasp on or-ganic, regional Earth-folk lifeway and consciousness. With the car as principal transport vehicle, TV was the main powder for accelerating the ongoing, exponential explosion of natural, whole self, family-community nurture, culture, mind, and being. The dilema is complex; but, the following description will suffice here.

To confess that the old, organic, Earth-folk ways of living are generally better than the modern ways, and then speak the details would be to delegitimatize our modern cults of education, sci-technics and commercialism, the arts, affluence, personality, nov-elty, curiosity, exploration, and cultural-creative-evolutionism, et al. It would disenfranchise modernism in particular and 'civiliza-tion' in general from the real, authentic, human journey. It would create a schizophrenic-like split between the transmuted, frag-mented, alienized being and reality (quasi egoself, quasi life and quasi world) we are forced by history, ethnocentricity and occupa-tions to live and the reality that is rightfully ours according to our evolved, genetic nature intergrown and interfunctioning with our coevolved, naturalistic, nurtureculture. Those of the keenest intel-lects and in positions to speak the sanctioned, romantic version of our lost reality can handle this madness better than the rest of us not so protected. But, on second thought, what have even the people got to lose upon receiving the human revelation? Is not one who lives in a cultural prison with the knowledge that he has been framed by history better off than one in that prison not knowing why he is there, except through some presumed, personal respon-sibility for one's life and fate! More precisely, is not the young performer, born, raised and generally confined in a circus-zoo, better off knowing that his selflifeworld is particularly artificialized (unnaturalized) and that a more natural (and therefore more free) selflifeworld is out there in the past and lingers suppressed within a false self! Many would answer that one should not know, so as to avoid a deep painful conflict. Of course, there is a crucial differ-ence between the above circus-zoo and the human circus-zoo of 'civilization'. The former can be bequeathed and perpetuated in due course. The latter is approaching a decimation of its alienized members and their alienized self, life and world.

The wilderness philosophers and deep environmentalists gen-

erally speak condescendingly of the people in the trenches embracing Earth-folk ways, people such as the readers of *Mother Earth
News;* for example, Paul Shepard has said "It is hard not to sympathize with them."[5] But, Earth-folk culture and lifeway have a long
record of holding good ground against urban 'civilization' for 5,000
years. To this point, take the Twentieth Century. The suburbs are
simply the flood planes for the raging river of history and ecological destruction; while rural communities are rooted in Earth-place,
time and human being, defiant in some measure to the thunder-
storms of sci-technic commercialism's media. Against the success
of Earth-folk culture and lifeway, what has a post-historic primitivism or an organic ecocentricism, as yet to be worked out and
agreed upon, to offer a humanity now in critical, possibly termi-
nal, condition?

In truth, it is more toward human and ecological redemption
to talk of substantially recovering the consciousness and lifeway of
the Neolithic, village farming-husbandry; precisely because this
entails talk of 5,000 year-old, Earth-folk culture and consciousness that substantially carry on the Neolithic village consciousness
and lifeway. The professional advocating post-historic primitivism
or organic ecocentricism encounters a problem here no matter his
choice regarding Paleolithic versus Neolithic paradigm. Such talk
extols the old in a culture that prerequisites novelty as predominant in any future world view. But, of the two paradigms, the
Neolithic presents the advocate with the greater problem. For, it is
more subversive and dangerous to talk about Earth-folk living because it is generally accessible to significant numbers, and challenges the legitimacy, profits and power of 'civilization'. Further, it
conjures up the back to the land faction of the 1960's counterculture movement, the limited success of which failed to leave a solid
impression as a viable alternative. Most important, to speak of a
movement back into Earth-folk family-community is to reveal the
breadth and depth of our ecological conflict: it is to oppose the
continued destruction of both Earthlifeworld's ecosystems *and*
naturalistic, human families and communities that together have

formed a 5,000 year old alliance against the destruction of ecosystems and human being, first by the city and then, modernism's
so-called progress. It is to polarize public awareness on the polar-
ized, spiritual-ontological, cold war (melting at points of ecological and spiritual conflict) of alien 'civilization' versus living nature
and her natural human being and naturalistic, traditional cultures
(folk countercultures). And yet, this public polarization is a necessary prerequisite to arresting the natural destruction that forbodes
leaving the Earth unlivable to both natural human being, and, as
this volume argues, the emerging alien being of sci-technic commercialism. We cannot hold much ground without seeing our
struggle as the 5,000 year old, spiritual-ontological, human, civil
war that it is–that our human soul is under siege.

## 4. The Limits of Wildernessism

An overlooked shortfall of the wilderness experience is that we
cannot reach and experience all of the untamed, wild living nature
some of us ultimately seek. Living nature includes the wild, untamed, social communities of the social mammals, such as the
wolf, elephant, monkeys, apes and Paleolithic humankind. This
part of the wilderness eludes or moves away from the wilderness
seeker. We owe some closer contacts with such social communities
of living nature to the Jane Goodalls of ethology and anthropology. But, even these observers, even when accepted as quasi members of a natural, social community, are denied a clear and complete revelation of living nature. The presence of alien outsiders
invariable alters wild social living nature. Would our behavior remain unchanged if aliens from outer space arrived to study us!

Our trips to experience the wilderness, to get close again to
the great Spirit, the Source, parallel in some measure the primordial forays from camp or tribal village to hunt game or gather
plant foods. It is as close as we can get to the spiritual-ontological
intercourse with wild, living nature we once experienced as natural humans before being alienized by 'civilization.' We travel out

of 'civilization', trying also to push it out of our mind and being so as to help living nature purify us during our reunion. The purification is partial because the Paleolithic family-community part of our evolved, natural human being is gone forever–infected by and gone extinct from the penetration and adulteration by alien 'civilization.' Further, it is temporary purification; for, when we return to the city, alien objects and phenomena reinvigorate the pathogenesis, alienism. Hence, the great importance of Earth-folk society as the 5,000 year old counterculture to city culture; the latest version of which is called bioregionalism. The family farm (sans television) is ten to one hundred times more in line with natural, human lifeway and consciousness than the contemporary city. For, the pathogenesis of alienism is much less successful in Earth-folk family-community rooted in a direct relationship with the land, in an ecoregion of the Earth. Even with television's thrust to rape the soul of every person, family and community on the Earth, Earth-folk lifeway, culture and consciousness still significantly survive as living nature's and human nature's compromise with 'civilizations' alienism. This compromise, last stand, human ground can be defended and preserved against the sci-technic, commercial, octopus-like, pathogenic illness revealed as alienism. It might even survive the foreboding decimation of 'civilization' s humanity. But not without a renurturalization that substantially recovers natural, human purpose and being-force authentically ours.

Another crucial element of our limited, human redemption is a family-community change in consciousness, recapturing the parts these play in human being within living nature's play of natural being. Individuals in larger, Western society, even if united in an electronic, global, mental alliance, cannot sustain, in a brotherhood of love for wild living nature, the degree and kind of relationship with living nature needed to save us and the rest of Earth-life from decimation. The depth and length of the authentic nature-human relationship, and any sustainable reconstruction of it, entails the family and community relationships with outer and inner living nature–with Earth-place, time and beings, with whole family and whole community nurture–nurtureculture. This is beyond society scattered throughout with a small minority of Thoreauvian-Muirean-Leopoldean individualists. The fresh wildernessism sprouting up from Thoreauism lacks a natural bent for searching and longing for our lost, natural, organic family-community paradigm of the soul. If Thoreau, after he developed his philosophy, had met and married a sympathetic woman, and had children, Thoreauism would be different; it might include organic family and community to complete the recovered, organic individual that amends (short of full correction) urban individualism. Part of this lack stems from the fact that wildernessism is partially reactionary therapy. This is not a flaw, something to be held against it. But, it is what wildernessism lacks that inherently restricts it from any hope that it may be *the* instrument of humanity's redemption or rescue. It remains a major instrument.

"In wildness is the preservation of the world."[6] We commonly misquote Thoreau replacing his "wildness" with our "wilderness." Thoreau realized what we commonly cannot: that wild, pure living nature dwells not just in wilderness lands but also within us, in our instincts, intuition and unconscious. It lies underneath the alienized 'second, human nature' of 'civilization'. Tragically, the closest words we have for the pure living nature within us are "instinct" and "the unconsciousness." This is because we have failed to break through to the revelation of purely natural socioculture–nurtureculture. Thoreau could only make a beginning at helping us to a closer, more complete reapprehension of the pure, wild living nature (natural human being) within us. This particular shortfall was his, as well as, ours. It stems largely from the egoism of the West, one of the cognitive strands spiraling up through our 'historical DNA'. The representations of reality the mind of this egoism presents are mistakenly thought to be legitimate in the absence of a fuller, nonegoistic, spiritual-ontological representation and manifestation of human nature's reality–natural human being (or beingness). *We think* from egoism's mind; *therefore we no longer are* (fully natural human being). Modern thought (of the intellect-egoself) separates us from both living nature and Her natural, whole, human self and human being. Over populated, 'civilized' humanity has destroyed most of both our pureness, and most of the pure, wild nature-human relationship, as well as most of the wilderness which harbors most of remaining wild, pure living nature. Therefore, arresting the contagious illness (pathogenesis) in our species that produces such behavior pathological to and within living nature is the other half, the first half, of human redemption and preservation. We are tempted toward a new *apparent* realization that the destruction of the human wilderness within humankind goes hand in hand with our destruction of the other wilderness; but, this would miss the crucial realization that the former surprisingly precedes, and is the source of, the latter. *Addressing the destruction of pure (wild) human nature within us (within our ailing mind-body organisms, ailing family and community) is where we start. For, we have departed from both our human part of living nature and the other, larger part; we are the aggressor, the megapathogenesis and illness within living nature. She has not violated and departed from us, infecting our human nature and being. The pathogenesis, instead, is within the human part of living nature. In order to preserve the other wilderness and its pure, wild living nature, we must rehabilitate and renaturalize the human wilderness (and its pure, unadulterated human nature—natural, unartificialized human being). We must clean up and arrest the pollution of the human wilderness before we can systemically clean up and arrest the pollution of the other wilderness.* This, of course, starts with the recovery of natural, authentic nurturing of the infant and child, which entails the recovery of natural, authentic family-community-ecosystem. Natural, unadulterated humans do not adulterate living nature; for, they are in an intimate relationship with her. *The environmental crisis is the destruction of the human nature (natural human being) part of living nature via the pathogenesis of human being by alien being emergent within natural human being: an alien debilitation spreading outward from human being to other natural beings and their Earthlifeworld (biosphere and ecosphere).*

We must realize that individuals (mostly individualists) arrive at wildernessism and ecocentricism in times of flight from their artificialized selflifeworld (alienism), which includes not just the artificialized, adulterated, mind-body egoself and its sci-technic, commercialized urban life and world, but also artificialized, adulterated, fragmented, diminished, surrogated, and debilitated family and community. If wildernessism and ecocentricism are the keys to the egoself's and the individualist's substantial redemption, to egoism's and individualism's substantial rectification within 'civilization'; then the key to whole human being's partial recovery and healing is the substantial reconstruction (recreation) of the natural, whole selflifeworld with its organic, whole family-(self) and community-(self) in substantial harmony with living nature's wilderness and pure, living nature herself without and within us.

## F. Humanity: Living Nature Become Self-conscious?

Many ecological thinkers hold the idea that we are living nature become self-conscious. But, our self-consciousness of 'civilization' in general and of modernism in particular has displaced our natural self (and self-consciousness) of the species with some historical, pseudo selves: (a) the intellect-egoself, (b) its collective 'civilization' self, and (c) very recently, an emerging global, electronic communications self alienized from living nature's Earthlifeworld (and whole human being or beingness within it) into artificial alienism (of alienized human being, self and lifeworld).

It is only when the self of self-consciousness is the natural one of our natural species (instead of the intellect-egoself, exemplified and advanced further, in careerism and consumerism )—only when it is the whole, organic triecologic self of bio-self, family-self and community-self, and their natural, organic lifeworld all enfolded within Earth's Earthlifeworld—that our self-consiousness can be a proud and loyal child of evolving living nature. Once humans transgress into artificial, alien 'civilization', they have lost their natural, authentic self and its self-consciousness. Historical force, like a man-made metal blade, performs its spiritual-ontological-meta-physical surgery on the natural, whole self of natural human being transmuting it traditionally into intellect-egoself and its collective 'nation-state' and 'civilization' self; and, just recently, into a proposed, emergent, global, communications-economic, alien-human self, which is accelerating the progressive removal of natural, human being and selflifeworld out of living nature into alien being and alien self, life and world, which, ultimately, cannot be anything but the finalization of our transgression and failure to perpetuate the human journey as a legitimate, viable primate species of Earth-life.

The cyclical failure of the first (above) two pseudo, human selves is witnessed through a human species-centric view (and for some, perhaps from a hermeneutical critique of recorded history). The (third) proposed, global human self (and 'global community') is blocked from attaining legitimacy and evolutionary sustainibilility by natural human being's constitution as evolved by evolution—natural being within (a) natural space (place of biologic-self, family and family-self, and community and com-munity-self, i.e., a bio-socio-eco-region) and within (b) natu-ral time of this natural, human selflifeworld. The artificial, alien elements (components) of modern, sci-technic, commercial 'civilization', e.g., modern transportation, communications and commercialism, eclipse and deny the alien scheme and dream because natural *Earthly* human being requires its natural *Earthly*, objects-phenomena constituents. In short, global conscious-ness in order to achieve ecological ends must include bioregionalism and the triecologic self (bio-self, family-self and community-self) lest it continues to become merely a corpo-rate-economic-consumerist, global pathogenesis spreading within evolved, living nature and natural human being.

# G. Economy, Ethics, Ecosystems, and Living Nature

The question, *Is the interdependence of living things we now call ecosystems a system of economic organizations or a moral community of mutual tolerance and aid?* springs from the embedded fragmentationalism of science and philosophy. In the limited scientific and philosophical sense (and mind) it is both collectively. Stepping free from the confines of the 'civilization' mind–free from science, philosophy, religion, and commercialism–this interdependence, this ecological relating, is an inherent feature of the organismicity of Earth's world (field) of living being. *The ecological world view must move beyond the interdependence of living things to the participation of all living things in the organismicity of Earthlifeworld. We are not autonomous entities that are merely interdependent in order to maintain our distinct identities and autonomies. This idea leaves us open to the doctrine of egoism pushed into the cult of personality–individual creativity pushed into egoself recreationism–of its debilitated, quasi 'self', ' life', and 'world'. No sci-technic, futuristic humanity is sustainable with Earthlifeworld due to the loss of parts of our evolved human being, our humanness, our human beingness. We must make this confession of loss to begin our recovery and preservation. Analogously, are the organs and parts of the organism (e.g., the modern mind) merely interdependent; or, are they involved in a participation in something deeper than economy, ethics and interdependence can account for and give us access back into!*

When the interdependence of living things is merely studied, with truths proposed for the sake of the scientific enterprise and scientific world view, then ecosystems is merely a science. When the interdependence is a given, preceding scientific study and part of a metaphysical world view, and a perception and consciousness, held even in naturalistic childhood, then it is a perception of reality later to be affirmed as metaphysical truth by philosophy and as methodological truth by science. The perception and experience of this wholeness of life and other aspects of reality precede science and philosophy. Science studies this interdependence from its primary goal to discover remaining unknowns, and secondarily toward the false hope of a unified knowledge of all unknowns. Philosophy contemplates this interdependence from one of two incompatible, primary goals (or merely from other secondary goals): (a) a complete metaphysics of Earthlife and 'civilization's' reality or (b) an unattainable, complete unified philosophy of the universe-cosmos–nature–human phenomena. What we lack at this late hour of the human journey is a science and a philosophy in synthesis together (as a philosence?) of Earthlifeworld, human inclusive, that overrides and subjugates science and philosophy of the universe-cosmos.

## H. Deep Ecology Versus Ecofeminism?

Our ecological quest need not labor under the false idea that we must choose between (a) the ecofeminist's type of caring grounded in a context of relationships that personally matter to the personal self (bio-self) and (b) the deep ecological caring that is grounded in the context of the web of life, in living nature's ecologic systems. The paradigm we must embrace must be the significantly reachable Neolithic that recovers some of the Pale-olithic. This mixed paradigm bears the lifeway (praxis) that modern 'civilization' must move back into as much as possible. It bears no intrinsic conflict between deep, ecological caring and ecofeminist caring, or any intrinsic clash of patriarchy and matriarchy. Both these conflicts arise from power-wealth-science-technology-commercialism ethos and dynamics of the city, state and empire themselves ('civilization'). These conflicts are festering thorns in the hands, feet and mind of humanity only after it leaves the folk, a'civilization', organic, family-community-ecoregion paradigm and lifeway. The point here is that in any natural, authentic eco-family-community retaining Neolithic and Paleolithic consciousness, these two types of caring are combined within a union of male and female caring within the family-community paradigm within liv-

ing nature; and, neither one gains dominance over the other. Together they are a part of a natural, human consciousness that provides both intimate, spiritual-ontological identification with family and community being, and with all Earthlifeworld beings as
these two natural beings, humanity's and living nature's world
(field) of being, participate together in Earthlifeworld's
organismicity—its play of life. Male and female being achieve
complimentary union in the whole human being that living nature evolves within its world of being. The characteristics of gender are always in complimentary union—sexually, psychologically,
socially, spiritual-ontologically—by decree of living nature's evolved
patterns and purpose. It is play by the rules or suffer some loss of
being—living nature's *human* being.

These two beings (or beingnesses), *human* and *living nature's*
*world (field)* of being, have a relationship distinct from that of the
genders. The human is the dependent child (one of millions) of
living nature and behaves her will until 'civilization' transmutes
humanity; whereupon, infantile or juvenile behavior erupts with
the ego's declaration of independence from the whole self—bio-self,
family-self and community-self. This separation of egoself from
whole human self and human being is at the same time a separation from living nature, since the paradigm of this triecologic self
is bequeathed for us (and for many other social mammals) by
evolved living nature, and since whole human being lives (and
bes) within living nature's Earth-world (field) of being.

Some ecofeminists and deep ecologists may seek to establish
that one of these two separations (a) the egoself from the family-
self and community-self and (b) the egoself from living nature
occurred first or is the root problem underlying ecological
destruction. However, when this first transmutation of Homo
sapiens occurs, the two separations are simultaneous. Toward this
conclusion, suppose it can be established that the idea to farm the
soil can only occur, or is more likely to occur, in the male mind.
This would have to overcome, among other objections, the
argument that it was a greater female understanding and

identification with living nature and growth that planted the first seeds along the stream bank that later became the alien-agrifields of the city. Nevertheless, regardless of the gender of the mind, the motivation is to feed growing family and community. The origin of the idea is gender neutral–aimed at feeding an increasing population of presumed, whole human beings. The original idea does not know of, nor has intention toward, the city's adulterating, fragmenting, transmuting alienism. This being the case, we are systematically innocent victims (of alien penetration and pathogenesis) in our fall from living nature (or Eden).

To whatever extent that sci-technic modernity and its separation of humankind from living nature are the culmination of patriarchy, then feminism's demand and desire to participate equally in modernity's crimes against living nature and human nature, to avoid being equally culpable, must be coupled with a cogent revolutionary or transformational platform to replace the destructive dynamics, only one of which is the male ego, with living nature's and human nature's growth-fulfillment-perpetuation. Whenever feminism forces males to attempt the nurturing that is distinctly assigned to women by (a) natural biology–the female body, hormones and mentality–and by (b) natural, Paleolithic, relational patterns of male and female growth and fulfillment, then, it merely compounds the human crimes against living nature and natural human being. Feminism's failure to be more than a bandage (at times, an exacerbate one) testifies to its failure and our failure to understand the depth to which the illness carried by 'civilization' has penetrated natural human being and spirit.

Patriarchy can be given a culpable role in the ongoing human separation from living nature without condemning it as the origin. City agriculture is a different matter. If the first ego or few egos to partially separate from family-community nature and being (by diverting some of the family-community harvest to create the royal wealth of the first city) were male; they did so on behalf of emergent power and wealth. This general, great departure from living nature was, in such a scenario, male initiated–unless the female gender in some hidden way stimulated this behavior. Males by virtue of their gender and physiologically based male ego are more likely responsible for the human '*triumph*' over living nature, the general departure-movement, as opposed to the *initial separation*. In short, there is, most likely, substantial female complicity in initial separation notwithstanding less complicity in the enterprise of city 'civilization'.

In 'civilization' the egoself secures its independence by alliances with other egoselves that effect the fragmentation and adulteration of family and community and the rest of what was whole, human selflifeworld dwelling within and participating with Earthlifeworld and its organismicity of life and species of beings.

# I. Science Penetrates Living Nature

Where living nature (Earthlifeworld) has been dragged into science and adulterated, she, and her human nature, have been violated. This is the initial act, general theme and scheme of 'civilization's' thoroughgoing, systematic violation of living nature that culminates in our present ecological crisis. After violation ecological scientists want to deny this part of living nature her place in the wild community of living nature. This lot of living nature is much like that of the woman that is raped and then stigmatized by her society that fosters rapists. And, scientists successfully get away with this disenfranchisement of the living nature that science has penetrated; because the penetration by science also infects and transmutes living nature with artificialism, with alienism. Unlike the woman raped and infected with VD, given treatment, maintaining her identity and human rights, science knows it has irreversibly removed the dignity, and altered the identity and nature of this living nature out of wild living nature. In fact, the pathogenesis-violation was for that very purpose, to '*civilize*' living nature. For the sake of modernity, we must deny deep within any removal of living nature's dignity and viability; and, instead, affirm our improvement of nature. It is the same for 'civilized' humanity. We have been dragged into science and 'civilization' by the historical force and growth of 'civilization', our human nature and being infected, adulterated, diminished–alienized from what it is rightfully ours by living nature's evolved patterns and purpose. The question is *what are we as victims of this alien illness going to do about our illness–this alien, spiritual-ontological pathogenesis– violating human nature and living nature, and thoroughly at large within our violated selflifeworld.* It is within the living nature of this alien 'criminal' disease, gaining enough power-wealth, to decimate human being and keep remnant, alienized humanity in the circus-zoo laboratory called 'civilization'. The ultimate end of this alienism is that of transmuting human being and Earthlifeworld into alien being and alien, artificial Earth-world, a world of decimated, declining life forms stranded outside of evolutionary life and its ecosystem (and on mars if possible).

Since this is an *alien* pathogenesis, a debilitation, within the natural, Homo sapiens part of living nature, we have no natural, human biological or social protections against it. Our natural defense is inadequate–our human being's self-consciousness, its abilty to identify the nature of its own species' being evolutionally created by living nature. (Our mind-body organisms, our biological-ego-selves are aware of the pathogenesis through symptoms but cannot identify and diagnose the illness as an alien one.) Our whole self and self-consciousness of our species human being can identify and distinguish human being from the other beings in living nature's world (field) of being. But, can it recognize *alien* being penetrating it? And, if it can, why didn't it repel the initial penetration(s) of this invasive alienism of 'civilization'? And, further, why has there been no school of thought as would seem to be the case, to propose that an alien invasion of humanity is taking place? One logical answer is that the first penetrant disabled, or circumvented our self-consciousness–which is another way of answering in the negative, i.e., that human being's self (and self-consciousness) has no defense against alien being.

Another answer to both forks of the question lies in the outlandish nature of the idea of aliens and alienism: that something from outside, that doesn't belong in living nature's Earthlifeworld, could be in that world! It is our 'second nature' belief *a priori*, that everything here must have been created or provided by the world and therefore natural and belonging to it. The key, then, may lie in the validity and legitimacy of our belief in a 'second nature;' because, if there is a 'second nature,' there is a second lifeworld–and self–that this 'second nature' applies to.

The belief in a 'second nature' and reality—a second selflifeworld—almost necessitates an *a priori* belief that our 'second nature' is better than our first; otherwise, we end up accepting the painful conflict entailed, namely, that our world is and has been degenerating away from a better selflifeworld for a very long time–and apparently inevitably so. *Almost necessitates* refers, for one, to the recurring, minority belief in primitivism, which envisions a Golden, paradisical or Eden-like world preceding a fall into a 'second nature' and selflifeworld of 'civilization'; wherein, the fall is usually to be amended in the future by some divine plan for redemption or some sci-technic, manifest destiny of utopian, human omnifisense beyond Earth's living nature and human nature. Primitivism, however, as it has been traditionally conceptualized portrays a humanity adulterated with Iron and Bronze Age trappings, and others from the city *per se*. The recent, post-historic primitivism of some ecological writers gets much closer to a purely natural humanity by embracing Paleolithic consciousness and lifeway. And yet, they will not abandon 'civilization' even in their purest theoretical or philisophical aspirations or visions of humanity. They and we are so deeply entwined into 'civilization' that it requires the deepest human revelation to free our visionary intelligence.

We are bogged down by 'civilized' humanity's rejection of a proposed, unadulterated, pure first nature unless set either in the context of Christianity and its promised heaven or in the context of reincarnation, or in a mystical, cloudy primitivism. Moreover, the first and 'second nature' referred to in these three cases does not propose alienism–which holds that we have been alienized through emergent, alien being from within; these merely propose a fall or decline in quality through alienation resulting from ideological, intellectual errors and misfortunes linked to a path toward eventual recovery, through a more enlightened and fortuitous cultural evolution, a divine intervention or reincarnation, in a future realm. Secular humanity, supported by science and philosophy, has been a long self-imposed confinement of itself and humanity; for, it intrinsically must hold any 'second nature' as superior, as evolving humanity upward through culture. Otherwise, science and 'civilization' is guilty of having violated and infected living nature rather than having improved upon it. It is this guilt that we cannot bear; and that progressively moves us to an ever-stronger embrace of sci-technics and commercialism as the instrument of a future redemption beyond living nature and our, first, pure human nature–with or without an alliance with a god that obscurely evolves with this anti-living nature evolution. The surprise to us, the hidden reality lying at our feet, waiting for us to kick the sediments off it, is that we the people (humanity) are not guilty; for we are not this 'civilization' we so artfully and desperately defend with reason and innovation as our natural, new, second, normative environment–selflifeworld. Correctly revealed and apprehended, our natural, authentic selflifeworld finds us human beings, in suppressed and oppressed reality, a part of Earth's living nature–naturally evolved human being as natural, human selflifeworld within Earthlifeworld. Further, there is no compromise on this matter; the compromises emerge only upon our gradual movement back into our natural, authentic human reality and being we have been unknowingly forsaking and deserting. It is not now credible to hold that 'civilization' is a part of Earth's, evolutionary, living nature–not after modernity's accelerating destruction of both Earth's species and ecosystems of life, and of organismic, human growth-fulfillment-perpetuation of our bio-self (organism)-family-community evolution.

There is no need to flee a guilt misperceived as our human

guilt. It is 'civilization' that *is guilty*; it is the pathogenesis within us. Moreover, natural human beings don't *choose* to be ill; rather, illness gets through our defenses, our immunity. True pathogeneses are a natural part of nature's life; but, 'civilization' is an *alien* illness. Living nature did not evolve and create 'civilizaton' as a part of her purpose, her lifebeing-force: evolving life. After four billion years, the alienism of 'civilization' emerged within living nature, from phenomena not within her evolved living nature (of natural systems) and purpose; and not within natural human being's natural features, patterns, purpose, and immunities; for, we are (sans 'civilization's' alienization of our being) a part of living nature within her. We as a species—the real, natural, human being part of us—are as innocent as the wild bear, wolf and monkey. The pathogenesis started within us and spread outward through living nature. Living nature has been infected by an alien pathogenesis within her human nature. Living nature and Homo sapiens are innocent; but, humanity is progressively destroying itself and living nature because of its *inability* of its infected intelligence and whole, human self—the inability of the transmutant intellect-egoself—to diagnose human and world problems as stemming from an alien illness within living nature's human nature—within her natural human being.

The alien illness, of course, has progressed further in some than in others. Since pathogeneses are harmful, even fatal to living being, this *alien* pathogenesis carries supreme evil with it, having over-powered our will to be ourselves, to be human being, to grow, fulfill and perpetuate our human being. We oppose 'civilization's' alien pathogens, pathogenesis, and its alienized people and institutions not because they are guilty of crimes and evil doing, but because they are unknowing carriers and agents of the alien pathogenesis that is evil and destructive; they are infected with alien being that debilitates living nature's human being and other Earth, species beings with contagious, debilitating alienism.

The people are victims; but, to the extent that they come to identify and fully realize that they have been felled by an alien illness, they have defenses and treatments stemming from this deep human revelation. And to the extent that evolved, living nature, without us and within us, has for 5,000 years, as resisting human nature–substantially resisted alienism through Earth-folk socioculture and significantly rescued fallen humanity by cyclically breaking 'civilization' downward toward Earth-folk lifeway and culture; to these extents we are not helpless victims. Living nature's human nature moves us to rectifications on the conscious and subconscious levels of our natural spirit and being. It is clear at this critical moment of the human journey that life will have to release the human revelation more generously, if it is her will and within her capacity, in order to redeem her human species and perpetuate the human journey through time.

The revelation of natural human culture and selflifeworld trapped in a mortal dialectical and spiritual-ontological struggle with alienism redefines good and evil, truth and falsehood, reality and illusion. It replaces the moral philosophies (of good and evil) by revealing the good to be one and the same as the injunctions of a recovered natural metaphysics and epistemology of our natural, authentic species human being and its natural selflifeworld and nurtureculture within Earthlifeworld. The real, the true, the good and the beautiful are rediscovered as one.

## J. Getting On-Line or Getting Off-Line?

Virtual reality began, in one sense, with the telephone as a kind of virtual or quasi conversation. When "talking on the phone," we can't see who we are speaking to–their facial expressions, body gestures (body language), their mental and spiritual-ontological demeanor and bearing–the realities essential to the person speaking and to what is spoken. We can't see the place, the environment of the conversation, which enframes species-language. This latter missing element, i.e., that we can't see where the conversation is occurring, informs us that it is occurring no where, which is to say that it is not *really* occurring in life, since the real phenomena of living beings occur in a place. Telephones, television, and computers surgically remove people from particular places where real conversations must occur; and, moreover, they remove us from one another, from activities, and relationships which require our presence in a particular place to be fully real. Phone talk occurs only in space and warped time. Indeed, all electronic communications separate time from space—time from place. Something audible and visual is occurring—as virtual or quasi phenomena. But, we, on the telephone line, or at the television screen or computer, cannot say where we are having the conversation, nor that we are where the televised or on-line phenomena is occurring. We stay in an alien place, and play an unprecedented, alien role in human relationship with others, and with time and space. We abandon Earth-time-place for warped, alien time-space. We have permitted electronic technology to be our *"virtual"*—adulterated, diminished, and transmuted—senses, perception, experience, intelligence, reality, life, and being. Human being is no longer real and natural—as evolved by and evolving with Earth's life—by virtue of its removal from real, natural conversations, relationships, activities, family, community, and particular, distinctive places. We are outside our context—the essential circumstances of human life and being. Human being is no longer on its distinctively evolved journey (of evolving growth) within Earth's evolution of living beings. Submitting to such surgery, human being is exploded into fragments darting about and tearing apart Earthlifeworld's time, place, and its ecoprocess-evolution of human being and other Earth beings. We become the true aliens to Earthlifeworld, destroying the beings in her world of beings, including our human being. Alien being emerges and grows within human selflifeworld: an alien, life form doomed to destroy its source-host form of life and realm, Earthlifeworld. Doomed to fail at being a lone form of life, a lone being.

We actually have been 'on-line' for a long time; and, as the hidden bill adds up, it is taken out of our *real* self, reality, human being and family-community. This bill, this spiritual-ontological deficit, if unchecked, will eventually overwhelm us, bankrupt our soul and decimate human being and spirit.

## K. Life and Non-Life as First, Primary Dichotomy and Perception

We may think that life and the inanimate together constitute one of the bi-polar, binary opposites and that 'civilization' and wilderness are, also, binary opposites. But, it is only the Earth's organismic, ecologic lifeworld, its evolutionary life, which pulls some of Earth's inanimate matter into participation in Earthlifeworld, that makes life and non-life two parts of Earth's whole Earthlifeworld. It is only in Earthlifeworld–not in Earth's core and not on the planets and in outer space (universe), that ecosphere and biosphere incorporate the inanimate into the process and systems of evolving life. We must recapture the perception and understanding that the emergence of Earthlifeworld and its meaning, purpose, beauty and consciousness set it apart and removed it from all that *universe-cosmos,* which lacked these. Before life emerged, destined for a system of evolutionary life achieving sentientness and consciousness, the inanimate did not require life because life was not here nor anywhere else in the reachable, empirically knowable universe–*the meaningful 'cosmos'* or realm of our Earthlifeworld. And, here on Earth, life does not require, is not dependent upon, the outer space universe in any meaningful way nor within any meaningful, future time. After all, if life at some time emerged on Mars, it had no meaningful destiny; it did not reach sentientness in a self-sustaining organismicity of evolving life. The rest of the Milky Way galaxy is moving and is removed so far away from the Earth's evolution of life in space and time that it is not a part of our natural human being's reality. It is part of an unreal, scientific enterprise, and meaningful to that alien, pathological mentality. It does not effect human reality–*natural* selflifeworld–nor will it in the meaningful future (the next few thousand or million years). A species struggling to endure a few more generations cannot afford the indulgent fantasies of outer cosmic meaning and reality, which, in fact, are illusions. *For a species to continue in the play of evolutionary life, it must identify with and practice its species reality that has its particular role, its particular place, time and being, in the play of life. We must do so without distraction, transgression and illusion, without so to speak forgetting our lines, our part, our boundaries, the purpose and destiny of our natural human being.* That life and death, the animate and the inanimate are found together is only meaningfully true–only in human reality– here in Earthlifeworld; the natural realities of which we must keep our selflifeworld and human being within. Otherwise, we are no longer secured in (and insured by) 4 billion years of evolving life.

Some argue that 'civilization' and wilderness are bi-polar opposites; that each is dependent upon and bound up with the other. [7] But, the wilderness, or wild living nature, is 4 billion years old; therefore, neither living nature nor her natural human species can be bound up with or dependent upon 'civilization'. 'Civilized' humanity cannot do without living nature and her wilderness; and, expressing our 'civilized', intellect-egoself and arrogance, we falsely believe she needs us also. But, in fact, we carry an alien pathogenesis within her that debilitates her and diminishes her richness, her ecosystems of life and being; just as this alienism originated in us and debilitates our natural human being.

Through science, philosophy, and religion, 'civilization' penetrates and effects a spiritual-ontological bleeding from the soul; in the midst of living nature, these change and denature living nature and pure, wild, evolved life. The perception, thought, actions, and enterprises of modern science, in particular, effects a reaction from sentient, conscious life–social primates, including indigenous pre'civilization' and preindustrial peoples, in particular. The result of this presence is manifested, alien perception, emotions and behavior. Hence, the scientist cannot understand real, authentic life; for, the scientist infects and transmutes that which he studies. And, if he resolves to preserve wild living nature or the wilderness, he is a moment too late; for, the preservation of evolved life *lies in*

*a natural way of living, not in an enterprise of artificial 'civilization'.*
Our way to a sustainable nature-human relationship–toward the
preservation of evolved life, which includes human preservation–is
to move back into that evolved life process and system which
created us and sustains us. This is a very slow movement into
healing and redemption. Still, our present course, of modernism
and progress, is the human exodus from Earthlife and its evolution.

## L. Cultured or Crabbed or Cancered

Gary Snyder says "There has been no wilderness without some
kind of human presence for several hundred thousand years. We
need a civilization that can live fully and creatively together with
wildness. We must start growing it right here in the New World."
This is a great dream for our heart and spirit. Indeed, Synder and
much of the deep ecological movement fully realize that "civilizations East and West have long been on a collision course with wild
living nature, and now the developed nations in particular have
the witless power [and intention] to destroy individual creatures,
whole species, whole processes, of the Earth." [8] Yet, he and they
will not or cannot renounce 'civilization' itself as intrinsically anti-
life or anti-human being. Their entwinement in it, and their addiction to it forces them to generally side-step two other truths:
(a) there has never been a *'civilization'* that has lived fully and creatively with wildness (wild living nature). Only the Paleolithic
(pre'civilization') part of our species lived fully within wildness
(wild pure living nature). Every 'civilization' that has appeared has
been destructive to (wild, pure) living nature and (wild, pure)
human being. And (b) the reason we have wilderness areas with no
kind of human presence (except the destructive presence of tour-
ism-recreationism) is because 'civilized' peoples pushed the indigenous peoples of the wilderness out of *their* wilderness–theirs because they as a people and socioculture were a natural, wild part of
the wild wilderness.

The dream of Snyder and some others in the ecological move-

ment–of an "old-new way" of living with living nature, or of building a "culture of the wilderness from within contemporary civilization"–is still infected with the *new*, 'civilization' part of us that has
penetrated, violated and transmuted our pure, wild natural human being, selflifeworld and nurtureculture (300,000 years old).
When North America (Snyder's "Turtle Island") was a land of
peoples fully participating with living nature's, wildness, it was a
land of no more than a few million indigenous people, free of the
alien adulterating ("developments") which has defined 'civilization': (a) overpopulation centrally manifested in cities, (b) complex, social organization and stratification, (c) oppression and dehumanization of the people ensnared by 'civilization', (d)
alphabeticized writing as the major part of (e) alienizing language,
(f) alienizing technology, (g) alienizing science, philosophy and
religion, (h) destruction of living nature's ecological life including
(i) destruction of family-community-ecosystem (bioregionalism)
evolution (growth-fulfillment-perpetuation).

Paradoxically, these magnificent visions of 'civilization's
transformation to ecological harmony with living nature–evolved
life is the source of all ethics, value, meaning, purpose and
beauty–is that the flaw of these visions lies in their greatness,
goodness and compasion–they are visions of the heart and soul
to redeem humanity–in reaction to the long domination and
oppression of the soul by the mind. Can such benevolent dreams
be found culpable? Hardly! But, they, nevertheless, are
unachievable; for, they grasp for something that humanity, so
progressed with the illness of 'civilization', cannot hope to
achieve. The web of sustainable, ecological communities and
societies we critically need cannot endure the presence of the
very essences of 'civilization' that are historically, consistently
pathogenic to living nature and her human being, as well as, to
any reconstruction-recovery (transformation) of living nature's
organismic world (field) of being that has evolved Earth-life
and our human species.

What most of those in the ecological movement need to grasp

is that it is not possible for a species to live for 300,000 years, and then, in the last five to ten thousand years suddenly discover its essences awaiting in 'civilization'. The essential, defining features of 'civilized' humans are only that—the *defining features of 'civilization' humanity*. The essences of our human species flowed through 300,000 years of time and being with us: *essence and existence are coevolved, coemerged (or coextensive); essence is not an attire we slip into upon the recent violation-transmutation we call 'civilization'*. We must not mistake the rediscovered *vision* of our lost, pure, uninfected human being (and beingness), the vision of what we were as evolved within living nature's organismic world of being, with the reality of what is feasible in the way of renaturalization and renurturalization—healing—of debilitated, transmuted, 'civilized' human being and living nature's living world of being. 'Civilization' is the pathogenesis that moves us out of both living nature's human being and her ecological world of life and being. Our crucial, feasible goal is to optimally preserve and recover the natural, authentic human selflifeworld (structure-function-dwelling) of our evolved natural being—preservation and repurification. We cannot be living nature's full, natural (Paleolithic) human beings again. But, after an age or some ages of human rectification and decimation, living nature may again carry some of us from out of 'civilization', together with Earth-folk a 'civilization' peoples, generally back into her time and being.

## M. From *Walden* to Nature-Human Village

"Life consists with wildness. The most alive is the wildest."[9] Most of Thoreau's followers have a vested need to overlook the painful realization that follows from this revolutionary principle: namely, that the opposite of wildness, of wilderness, of living nature—'civilization'— is the least alive, and that 'civilization's 'civilized' humans are diminished of life, are life in decay. To tame life is to partially kill, with chronic illness. This entailed realization has been comfortably overlooked by most readers. How do we justify, more deeply than on immediate, entrenched, pragmatic grounds, a post-historic primitivism or any ecocentricism that drags the culpable essences of 'civilization', which have infected and critically debilitated evolved life and human being, into either a recovered prehistoric paradigm or a postmodern paradigm of the human being-nature relationship proposed to redeem us from our tragic and possibly fatal transgression of natural human being and evolved, living nature!

Thoreau prayed "for such inward experience as will make living nature significant."[10] It is ironic that this great *philosopher of living nature or wilderness* liked to call himself "a writer of journals;" for, *writing itself is artificial,* an act of historic consciousness and lifeway; it began at Sumer, within 'civilization', outside the wilderness, outside (wild) living nature and her (wild) human nature. Thus, Thoreau's "need for such inward experience as will make nature significant" was partially, in reality, a need to counter, to *heal* his natural being from the writing obsession infecting his natural being. *Since writing is blatantly unnatural, how can we reconcile the professional writer with the spiritual quest into living nature? And, for that matter, how do we reconcile the computerization of writing, communication, conciousness, perception, and our lives with the ecological movement! In principle, and in long-term praxis, there is no reconciliation of increasing artificialization of our selflifeworld with a movement to preserve Earthlifeworld (wild, pure living nature) and our human nature and being within it.* Only the writer, among all the mammalian beings, self-secludes his egoself away from the normative life of his species being for long periods of time to scribble expressions to his egoself and other egoselves of their quasi, transmuted lifeworld; and, only the alien city accepts and depends on this act of alienized being. In the above quotation, "inward experience" refers to the inward intellect-egoself of the American (and Western) individualist—rather than the whole, human self. Thoreauism's and wildernessism's limitation is that the wild living nature (to which Thoreau prayed) is not all of wild, pure living nature. The biologic-self (organism), the family (part of the self and being) and the community (part of the self and being) of

Paleolithic (natural) humanity are also parts of pure living nature. Thoreau knew our biological organisms (bio-selves) were part of wild living nature. But, the latter two social and spiritual-ontological parts of human nature were generally beyond his perspective; and rather completely left out of our perspective and conception of living nature. Thoreau and most of us cannot make it back to a Earth-human village, though the movement closer to that home village is crucial.

Thoreau spent most of his allotted hours running back and forth between his alphabetized, written expressions and living nature's resisting enclaves, of woodlands, meadows and rivers. He clearly had been converted from transcendentalist to linguistic philosopher, as much as to, philosopher of nature. He emerged, from his pursuit of truth through the study of Greek and other classics, to a prose that asked nature's inspirations upon him to create literature that speaks of nature's truths. Thoreau, like many of the contemporary wilderness philosophers and deep ecologists that admire him, was trapped within the cult of language (and particularly the cult of writing), which holds that language is the essence of the human. The susceptibility to obsession and ideological addiction, from which the intellectual cult (the cult of reason) singles out any of our essences as *the human essence*, is characteristic of 'civilized' people and the West in particular, and certainly also contributes to the genius or madness question. Thus, so much of Thoreau's identity and functionality involved writing that he could not break free of his profession as "writer of journals" and books, nor the exhaltation of literature it generally entails. Thoreau's effort "to live deep . . . to put to rout all that was not life . . . to cut a broad swath, to drive life into a corner"[11] would have to spare literature; for, the literary community was his primary community and literary expression was his primary expression. These secondary constructs Thoreau was forced, by combined personal and cultural circumstance, to generally surrogate for the real, natural family-community and the expressions (including natural language) of these that are authentic to and primary in our species human being. As an unmarried, professional writer, there were insufficient natural, authentic primaries of human being to balance, the surrogational secondaries, much less to coordinate these to an spiritual-ontological clearing and field of (bio)self-family-community, Earth-rooted human being. We cannot pass judgement on Thoreau; for, like all 'civilized' souls he was born into not only natural forces, but also artificial forces (anti-life, non-living nature, and alien forces), that manifest the adulteration and transmutation called "a civilized being." It was his misfortune and that of most of us to lack enough natural, authentic family-community (or a prophetic friend) to help separate himself from writing enough to subordinate it (at least for long periods) to his pursuit of natural human being. He lacked a clear vision of the wholeness of the lost human being (beingness) he pursued. An obsessive, intimate relationship with wilderness or Mother Nature is the soul's substitute for living nature's family-community, *nurturecultural* part of natural human being, lost to the fever and swellings, the pathogenesis of nurture and culture, and carried by the sweeping, outward spread of pathogenic, historical force into wholesome lifebeing-force, human selflifeworld and Earthlifeworld. With a more natural, social make up, he might have been able to behold the ultimate vision of lived, natural human reality and being. A small minority of frontiersmen, fur traders, gold prospectors and the like, married into Native American socioculture and a few of these generally accomplished what is commonly believed to be impossible: a move backward in historical time. As it was, Thoreau was confined to his stunning genius of getting nature to speak some of her truths about her and humanity through Thoreau. With more luck, Thoreau might have had a deeper destiny; he might have been America's much needed prophet. As it turned out, Thoreau's shortfall is to the great benefit of American literature, the language as human essence doctrine, wilderness philosophy, and the natural environment.

The need to give nature a mouth and find words tumbling out, to give her a brain and language, smacks of Western religion's need to create God in man's image. Does this call to nature to speak a verbal language to us, to literally get into our language–and to get ourselves to relearn to listen, hear and speak a natural language–spring, in part, from a need to humanize living nature, to create her in our human image? And, if so, is it a legitimate need of natural human being, or a hidden retention of Western anthropocentricism and a reflection of the contemporary, Western, linguistic language orientation toward truth and reality? To be sure, to organicize our language works toward overthrowing the Judeo-Christian-Cartesian-Mechanistic paradigm or world view; but, surely, the switch in language would follow (or be simultaneous with) the switch in world view, since we speak of how we perceive and think. Our call for organic language and nature fabling is well grounded in a crucial need, for a recovered, organic, world view, and more deeply grounded in living nature and her Paleolithic, natural speech. But, we must not mistake it for the solution, which is to renaturalize and renurturalize human being, its selflifeworld, and nuturome. Ultimately, people will only talk organically after they see, perceive, think, feel and live organically. Our main obstacle is our deep, ethnocentric egoistic need to view 'civilization' as a part of living nature: a need to have it inducted into evolution's hall of fame and legitimacy. This 'civilization'-inductionism is our main obstacle. This violational, illusional, exploitive pushing of 'civilization' into the story of life–the evolution of life–this is the rape of all evolving life, which includes and even starts with self-violation. It flies in the face of 'civilization's' general, inherent destructivity of living nature and our human being within it. Our redemption requires a move toward living nature's mouth *and* her arms, teats and loving nurture. For, if we are her children, she gave us both a natural, pure language to speak the voice of our natural being to us and to her, *and* a natural culture to manifest, fulfill and perpetuate our natural human being. To recover natural nurtureculture is to recover its trappings of language, customs, rituals, patterns and lifeways through regrown whole self, being and consciousness. There is no key feature: growth is *organismic,* not centrally linear, rational, linguistic or technic. Human being unfolds–grows and fulfills its nature and destiny–from bequeathed natural genes and natural culture, nurturome, unfolding and evolving in a natural union that carries all our essences, such as natural relations, feelings, rituals, tools and language. We must break free of the key essence syndrome, as well as, the 'civilization'-inductionism syndrome. The grasping for a key, human characteristic or essence exposes again our 'civilized' need–and our 'civilization'centricism's need–*to separate ourselves from or elevate ourselves above or place ourselves at the top of evolved, living nature.* This assertion is made notwithstanding the fact that the key essence thesis is also used *to overly distinguish* humanity from other species. The problem occurs when the presumed key essence is exalted, which is nearly always the case; for, this generally leads to the glorification of humankind as the jewel of evolution, which is a dressed up, postmodern moderation of modernism's human arrogance. Since all of our essences, e.g., tool use, language, etc., can be found in simpler forms in some other social mammals (e.g. dolphins, chimpanzees and Homo erectus), there is, in truth, no basis for separating Paleolithic humanity from living nature; and, the glorification is strained and without merit, notwithstanding our greater *appreciation* of our human being (and that of other species for their own being) as a part of species self-preservation. It is strained (more precisely meritless) on the same grounds that talkative–or educated––people are not more fully human; nor do they experience more human being than quiet–or preliterate–people, respectively.

The subject of 'civilized' education, brings us back to human authenticism's view that both separation and exaltation of humanity stems out of 'civilization'centricism, and particularly out of 'civilization's exaltation. 'Civilization' and civilized humans are simply so inferior in too many ways, to permit us a normative or superior status in respect to natural, authentic human being. Our acquisition of a great quantity of knowledge of Earthlifeworld and the universe cannot be vindicated because it cannot be separated from our lack of ecological, spiritual-ontologicality and spiritual intelligence, nor from the resulting, accelerating destruction of human selflifeworld and Earthlifeworld, which made this arrogant, destructive knowledge possible. *Modern 'civilization' wants to have the cake of mind, while devouring the grains of being that make such extravagance (of intellect-egoism) possible. Such is the modern cannibalism, devouring our human being, soul, and Earth-life.*

The view, derived from human authenticism herein, is that the language as human essence thesis or doctrine of our time stems both from the embedded anthropocentricism of Western culture, and, more important, from the more deeply embedded and more pervasive 'civilization'centricism that is our ultimate obstacle to apprehending natural human being, its reality and selflifeworld. Language exaltation is part of an attempt to recreate living nature in our 'civilized' image; and, further, in its double role, it is a rationalization for the communications revolution which explodes into the face, consciousness and culture of beseiged, natural human self, human being, as well as, the body and spirit of Earthlifeworld. The language as human essence thesis is startlingly revealed as a two-headed serpent of 'civilization' blocking the path to a post-historic primitivism or any ecocentricism, whether Paleolithic or Neolithic, and blocking any ecologically sustainable consciousness.

Thoreau tells us, "all nature will *fable*" if we will but let her speak. But, we easily overlook or forget that neither the wilderness nor the wildness Thoreau apprehended is all of living nature. If we and Thoreau could but have lived two weeks in a hunting-gathering village or camp! "If I am overflowing with life, am rich in experience for which I lack expression, then nature will be my language full of poetry,—all nature will *fable* and every natural phenomena" (of the wilderness and wild, biological senses, instinct, and intuition) "be a myth."[12] But, just as the wildness in wilderness and in our organisms is only part of living nature, and its expressions are only part of her expressions; living nature is only language in a partial sense. In the full sense, in human reality within living nature's reality, *living nature is living being*, of which expression and verbal expression are major parts. The literate professional cannot be allowed to drag all the expressions, the phenomena, of living nature into the phenomenon of language for the sake of his cult of language and/or literature. This is close to intellectual hocus pocus. In reality, it is the other way around: language is one particular phenomenon ("expression") within living nature's phenomena ("expressions"). Thoreau gets as close as a person of literature can get: "he is richest who has most use for nature as *raw material* of tropes and symbols with which *to describe his life*."[13] This sentence reveals the utilitarian and anthopocentric parts of Thoreau ("raw materials") and the rank individualism ("to describe his life") that eclipse the whole, tri-ecologic self of bio-self, family-self and community-self; the parts that were also in conflict with (and obstacles to) the Thoreau that sought Indian wisdom.

Living nature, indeed, can go beyond fables; she can *show* us the way, and carry us along when we jump into what is left of pure living nature and human nature. Do we not learn—grow and fulfill—at home and school from both being told and being shown how the achievement-nurturement—the growth-nurturing—happens, and are swept along into it? Our obsession with and addiction to adulterated, swollen language, knowledge and culture is our limitation, our obstacle to living the truth and reality of living nature's authentic human being. The swelling is that from pathogenesis with artificial, alien elements-phenomena, including abstraction itself. The swollen intellect-egoself infects natural language in its role within natural nurtureculture, nurtureculture itself, selflifeworld and Earthlifeworld.

Living nature can both fable and show; and more, she can lead, can carry us along in her flow of growth-fulfillment-perpetuation, if we will but recover her unadulterated, unobstacled nurtureculture, our nurturome, within which, intergrown with genes, she carries us along in the human creek of Earth's watershed to her sea, and sky, and land. Beyond Thoreau's "Simplify, simplify, simplify." awaits, *Purify, purify, purify*.

Thoreau asks, "where is the literature which gives expression to nature?"[14] But, living nature has never needed literature to express herself nor to express the human part of her. Only living nature can express nature's living being, and does so through each and every species' experience of being, including the audibles of being, such as bird song, rudimentary, primate speech and Paleolithic, human speech still free from adulterations (like abstraction). To natural humanity literature (and abstraction) is unknown. It began at Sumer, a departure from living nature's human expressions of natural human being, one of which is preliterate speech. The written word first began to speak and still speaks for 'civilization's artificial alienism that infects the writer and the reader. It is only in spite of its fall from natural, human expression, only in spite of itself as debilitated, infected, human expression, that literature–as an artificial obstruction and adulteration of natural language–sometimes still speaks some small truths for living nature. Any literature that achieves the expression of some of our natural human being still alive, still being experienced, in that way and to that degree, gives expression to living nature. Great literature releases expressions of living nature's human being resisting the grip of 'civilization's alienization. This marks great literature; all other is worse than useless and meaningless; it is falsehood expressing the illusions of 'civilization'.

Thoreau's effort, and that of the language as human essence school, to create a literature that speaks for nature reveals the capture of Thoreau and the school by the cult of literature, and their failure to separate literature and language–both as living nature, and natural, Paleolithic human beings do–by preceding and excluding literature. To drag oral folklore into literature, disguising discontinuity as continuity for the sake of the cult of literature is a newly dressed-up, arrogant 'civilization'centricism. The professional, 'civilization'centric thinkers carry a vested need to pull primordial speech, lore and other authentics of our species into the bag of 'civilization's synthesized transmutation, in order to avoid seeing or confessing the full nature of 'civilization's transgression from and violation of living nature and natural human being. It cannot be overstated that literature emerged within 'civilization' and belongs to it as a rationalization of it. Hence, how can literature speak for living nature which declined to create and include it? Great literature—naturalized literature—releases *some* utterances of living nature and her human nature (despite its being an artificial construct) but, it cannot be a language, a voice or an expression of living nature after she has excluded it from our natural, species human being.

Natural human being requires *language* as one fundamental expression of it; but, it did not for 300,000 to 1,000,000 years require literature, and still excludes it, notwithstanding our deep denial. Only adulterated, infected, alienized human being of 'civilization' requires literature as one of its rationalizations, as one part of its fallen, alienized culture, intellect-egoself, and lifeway. Naturalized literature (secondary to naturalized language) remains only one of our links to the natural, whole human being we lost when infected by 'civilization's alienism.

# N. Epilogue

At the bottom of the environmental struggle and debate—and of human welfare itself—lies the issue of *what we are as human beings* and *why we are*. Modern 'civilization' accelerates our movement out of *what we are* and *why we are*, away from collective self-knowledge, a movement that begins with the emergence of 'civilization'. Yet, the most profound and earnest in the environmental movement propose to somehow retain the essences of modern 'civilization' in particular and 'civilization' in general, which are destroying Earthlifeworld and our natural, human nature and being within it. This stems from the first and most fundamental misunderstanding of the 'civilized' mind: that we can have it—lifeworld—both ways. Namely, that we can carry anti-thesis within thesis, the power-wealth-domination ethos and order within the organis-

mic order of evolved Earthlife, within which lies the human species. However, once we have come to receive the human revelation
of what we are and why we are, we apprehend that 'civilization' is
a pathogenesis within living nature's Earthlifeworld, and is emergent from within 'civilized' humanity. The extent to which we can
decrease the illness within us, within human selflifeworld, will
determine both the future welfare and perpetuation of the human
species and the other species of Earthlifeworld.

\* \* \*

During the Twentieth Century, the modern mind has increasingly dug into the idea that humans can or do exist independently
from the other species, from Earth's world of interrelational, ecological beings. From *Flash Gordon* to *Star Trek* to colonizing Mars:
all these fantacies proceed with the false belief in independent,
lone, human being (and beingness). However, in fact, all life forms
exist through their participation in an evolving process (system) of
life; pre'civilization' and a'civilization' Earth-folk cultures are quite
aware of this. Our incremental, slipping away from this primordial, instinctive understanding is our slipping out of Earth-life's
world of species beings and out of our *human* being within it.

\* \* \*

A wide, ecological identification with Earthlifeworld can be
achieved only through a spiritual rebirth of Western humanity's
alienized human being—not simply individual consciousness
expansion and identification with all of living nature. The rebirth
is of the whole self, its family(-self), community(-self), the natural
selflifeworld and nurtureculture, our nurturome, of these that are
*within* living nature; after all, these are evolutionarily created by
living nature, are intergrown, interfunctional, interfulfillmental,
and interpetuational with living nature's ecosystems of growth-
fulfillment-perpetuation of life and being. This organismicity of

human selflifeworld and its nurtureculture has its natural boundaries and must remain separated (generally) from the adulterations and violations of alien, intellect-egoself and alien, city, lifeworld culture of sci-technics and commercialism. Whenever they have co-mingled, the alienism acts as a pathogenesis within natural, human authenticism impairing the health of human life, being and consciousness, which then spreads its malignant-like pathogenesis to other portions of Earthlifeworld (the biosphere and web of life).

A wider, natural identification must respring from living nature–must be reborn, renurtured, regrown, and refulfilled within reborn human being and its natural, authentic nurtureculture, once the illness of alienism is arrested. Ironically, the rebirth (and renurturalization and renaturalization) is not natural but occurs only as a revelatory reaction to an alien (non-natural) illness. Our recovery, our redemption–our freedom to be human beings–will naturally take two or three generations to be a substantial one. As usual our children lead us as we help guide them out of the alien, artificial jungle.

\* \* \*

Most everyone in the deep ecology movement now finds modern 'civilization's activities to be the general culprit of our ecological crisis. But, even the most notable critics will go to any length, resort to any rationalization to save the legitimacy of science, philosophy, and 'civilization' themselves, and their essences, from this guilty finding. But, we need to find 'civilization' and its characteristic enterprises guilty, and sentence them to the indefinite public service of making amends to and restorations of natural human being and nature.

\* \* \*

The environmental movement must clean up the household of mind and perception as part of cleaning up the selflifeworld of

'civilization'. As we make the movement homeward, some embedded pieces of alien furniture must be scrapped. Very notably: (a)
we must toss outer space back out into outer space, out of our
sight, mind, being, and selflifeworld except as the firmament it
was before artificial, alien inventions (telescope, etc.) pulled outer
space down, before the sky fell in upon us, before it invaded the
Earth and occupied our human-Earthtime, being, spirit, lifeways,
purpose, and evolutionary destiny. And (b), we must toss inner
space back into its real place where it was before artificial, alien
inventions exploded it outward through our natural mind, culture
and natural selflifeworld, invading our time and being, and our
evolving Earthlife and being. Pragmatically speaking, *we* can subjugate *them* (machines)–reversing the modern situation so destructive of evolved humanness and Earthlife. We can put them out on
the back porch so to speak. We must do this to effect some healing
and recovery from the illness of 'civilization', alienism.

* * *

To affirm that humans were once a completely natural species
is to affirm completely natural, primordial culture. Centuries of
recorded encounters with pristine peoples affirm natural culture,
including language. Moreover, contemporary ethology and anthropology also affirm natural culture. Language presents no special
problem for the affirmation of natural humanity. It is only educated people that spend time trying to distinguish true statements
from false ones. Once, all utterance was true to reality–a part of
reality (as much so as laughter, moans, humming, dancing, and
chanting). Language takes flight from such sounds.

* * *

Nearly all wilderness philosophers and deep ecologists want
to have the vast cosmos as part of the wilderness, and, also, 'civilization' as part of the wilderness. This leaves nothing that is not

part of the wilderness: if everything is wilderness, then it is not endangered, and extinctions are merely part of some all-inclusive, transformational flux of living nature. Such thinkers may reply that it is wildness (wild living nature) in the wilderness and in us that needs to be conserved or preserved. But, how can wild living nature, harbored by the inner and outer wilderness, be preserved without preserving that which harbors it? In short, if everything is natural, and of the wilderness, then wilderness philosophy is a rationalization representing a new, fresher, intellectual workout for these professionals of artificial, 'civilized' cognition and consciousness. These soldiers, seemingly, have yet to identify and distinguish the enemy in a Vietnamish struggle, with the Earth's natural environment and wild, pure living nature within and without us as the disputed territory. In truth, the enemy is the intellect-egoself's centralized illegitimate power and wealth. And, the worthy cause is the defense of autonomous, whole, organic self, family, and community of an ecoregion.

Science and philosophy, or the educated thought of 'civilization', have failed to realize that there are two separate, distinctly and fundamentally different spheres of nature: (a) the universe and (b) Earthlifeworld. Life's emergence was the Big Bang creating Earth-life cosmos. Any other *explosions of life* are unreachable and alien to our evolved Earthlife, and incapable of real, evolved, healthy interaction with Earthlife. This second 'bang'—the creative explosion of evolutionary life on Earth—created the only cosmos to which we and evolved life belong, and in which there lies any significant reality. The failure to separate life—the phenomenon of evolved life and its world—(a form of which is what we are and the world of which we participate in) from non-life (its realm and phenomenal universe), which Earthlifeworld departed from four billion years ago, as a new, fundamentally different creation leading to sentientness, consciousness, meaning, purpose, beauty, value, and lifebeing—as the story of being's journey through time: this is a major failure of us as Earth beings of 'civilization'. Our failure is our weak or absent embrace of our Earthlifeworld as the realm we belong to-with, the phenomenon within which we participate, and which bears us forth in life's evolutionary creation. The reach for the sky beyond Earthlifeworld as a realm of methodological scrutiny and wonder to the extent of the actual grasping and pulling of it down crashing into Earthlifeworld is an act of 'civilization' and its 'civilized' people's self-destructivity. The sky really is falling, but only because we have fallen from our natural human being, its consciousness, and the embracements of natural consciousness authentic and essential to our human being.

If it is unhealthy to daily break away our doors of perception with hallucinogenics, why do this with sci-technics and commercialism?

# CHAPTER X

# BRIEF APPLICATIONS OF HUMAN AUTHENTICISM

## 1. Search for Nutrients Vs. Search for Discovery

Is LIFE, INHERENTLY, a searching journey for adventure-discovery and subjugation-exploitation toward the ends and purposes of power-wealth, its ethos, and empires? Or is life a journey of searching for nourishment, i.e., for the cyclical, seasonal and generational nutrients–the material-biological-social-spiritual-ontological-nurturecultural ements-phenomena–that nourish the growth-fulfillment and perpetuation of what we are evolved, ecological to be, that nourish our natural being, spirit and reality! These two, (a) the search for adventure-discovery and (b) the search for human nourishment are similar in appearance; but, understanding the difference between the two is crucial and vital for the preservation of both 'civilization' humankind and all humankind. The search for objects-phenomena that are nourishing to the growth-fulfillment-perpetuation of human selflifeworld is natural and normal throughout the sphere of living organisms, throughout life itself. By contrast, the adventure-discovery of most historic, 'civilization' sociocultures has been expeditions and projects of searching for power-wealth, for commodities and knowledge highly valued in the adventurers' home society and for lands and peoples to be slaughtered, conquered, enslaved, exploited, and subjugated.

This historical trangression from the normal, natural search for human nourishment, a search fitting and authentic to our natural species' being and spirit (to our nature), has more recently been joined by another transgression coming from a newer type of adventure-discovery, i.e., the sci-technic adventure-discovery that likewise lacks as its central purpose the search for human nourishment, elements-phenomena that nourish people as human beings, families, communities, social and sociocultural growth, perpetuation and fulfillment. What has come to flourish in the twentieth century is an adventure-exploration-discovery ethic with the primary purpose of discovering upon anything whatsoever that is discovered, which is then to be examined for the primary purpose and enjoyment of studying the previously unknown itself; the secondary purpose is the happenstance discovery of spinoff elements that are found to be of some use within sci-technic sociocultures.

This is not a case of insignificant distinction. On the contrary, human nourishment—our natural, authentic, human purpose— should be the central, generating force directing human activity; it is instead a spinoff consideration of our adventure-exploration-discovery spirit-ethic. This is nothing less than the subordination of human welfare to what is, in effect, the ethics of adventure for the sake of adventure, exploration for the sake of exploration and discovery for the sake of discovery.

In fact, human welfare within 'civilization' sociocultures has for thousands of years been sacrificed, through the adventure-discovery and subjugation-exploitation spirit-ethics of the ruling classes' power-wealth spirit-ethos. More recently 'civilization's sacrifice of human welfare, purpose, being, and spirit has accelerated due to the acceleration of hi-sci-technicism; the sacrifice, now, seems, less solely for power and wealth, and more for the sake of these ethics of sci-technicism. As the directors of the power-wealth institutions sense that the wealth of the Earth is being destroyed, their purposes modify to acknowledge two basic facts: (a) that the mechanismic destruction of the Earth and human selflifeworld can be studied–dissected and adulterated almost as indefinitely (until the scientists and, perhaps, everyone are destroyed) as is permitted by their nearly indefinite complexities. And, (b) that both outer space and the disintegrating quasi lifeworld of 'civilization' sociocultures can also be studied that far.

They are unaware that the human revelation is dawning and will spread a renunciation of 'civilization's tradition of human sacrifice, earlier masqueraded as the will of some god interpreted by a king, and more recently masqueraded as human "progress". They are unaware that ultimately we all, in order to survive, will have to rediscover and recover much of our real, natural selflifeworld with its human nature, being, identity, spirit and reality, authentic to us and right for us.

The expenditure of resources, both monetary and human resources, into the alien, and the alien, artificial realms of life cannot be justified until that time is reached when a far more beneficial search for lost, human nutrients and beneficials is farther along. Until this legitimate search for and rediscovery of human nourishment lost by 'civilization' is completely undertaken, the illegitimate counterpart activities of science and technology cannot be justified by a human species now threatened by megascourges, decimation and extinction itself. It cannot be tolerated by our species that lives in an Earthlifeworld having its ability to support human and other life forms increasingly threatened.

After all, by way of analogy, would a band of hunters choose to explore a new region hoping to find game there, when they can be certain that the game they need can be found in another area that is more easily accessible! And, would a group of farmers squander their time and seed to till unfamiliar soil regarded unlikely to produce, when a proven harvest field is readily available to them! The expenditure of human time, labor and resources for the purpose of satisfying curiosity itself or challenge itself is a luxury that can only be afforded when all legitimate human needs are being met with mere, professional desires are subordinated.

Knowledge originally was a part of intelligence, directly related to the human needs for that which nourishes our growth, perpetuation, and fulfillment. It was, until, 'civilization', transmuted knowledge and it became exclusive–attainable and belonging only to the literate, the educated urbanite, the royal or ruling class, the upper class and recently, the middle class: until it became interlinked with the tools and institutions of power-wealth. Natural knowledge–nonliterate, nonurban, with sufficiency-growth-fulfillment as its purpose–was delegitimatized by those enveloped within the power-wealth, 'civilization' lifeway and thought-mode.

Recently, knowledge has acquired its own rationalization–knowledge for the sake of knowledge. This last knowledge is destructive to those utilizing it and those subjected to its cold separation from common sense and wisdom. Thus, knowledge has proceded on a deteriorating continuum, starting from intelligence, i.e., natural, beneficial adaptation to natural environment and natural camp and village sociocultural environment; crossing into adaptation to the alien, artificial, problematic conditions of 'civilization' socioculture; proceding into progressively greater such problems of Western sci-technic socioculture adapted to through intellect's reason–rationalism. From this course, human intelligence is transmuted into a rational, educated stupidity that unlearns what the nature of human being is and how it is fulfilled and perpetuated: transmutant intellect-egoself employs artificial change and 're-creation' to dehumanize and transmogrify our nature with adulterating, alien artificials that progressively debilitate our human being, spirit, our families, communities, 'civilization' sociocultures, and ultimately, all or most of humankind.

# 2. Motherhood I, Mother Nature, and the Aliens

We have met the aliens, and they are us. Many of us realize the urgent need for sci-technic humanity to cultivate a keener understanding and a finer appreciation of our first mother, her motherhood, and a close relationship with her. We have become alienated from Mother I, and harmfully deprived of motherhood I. Increasingly, we are aliens to mother Earth and her ecologic lifeworld.

Sure, to us it is an axiom that life inherently bears tragedy. But, one of our ironic tragedies is not within the axiom, and must be renounced and rectified. Ironically, the very forces directing our sociocultural evolution are, tragically, at the same time diminishing the two motherhoods that are essential to our human spirit, being and our humanity—as we have known, recorded and nurtured the human journey; as we have played out its stories. Humankind cannot control the pathogenesis and diseases—psychological, social and cultural—that decay our minds, social units, and even our vital humanness itself without preserving our twofold motherhood itself.

We of Homo sapiens sapiens (about 50,000 years old) are indeed born and mothered twice: first (a) by Mother Nature for our human species-being and (b) by a mother-person for our family and social human being, identity, and personhood. We are born and nurtured both of (a) mother Earth, recipients of her natural nourishment—biological, social-ontological-spiritual, and (b) of a personal, biologic mother that nurtures our body, mind and spirit. Ultimately, our two mothers are revealed to us as unified, as natural, authentic, human motherhood. The creation they in union create is carried onward by them, the family, and community. "It takes a village." And more: we are in the circle of Earth-life. But increasingly we adventure out of the circle.

Why do we go to living nature for her brief presence, or for a visit or stay? Primarily (and largely subliminally) we are seeking a closeness to our primordial mother, a relationship that existed before she and we were edited for the uses and purposes of 'civilization'. We find her timelessly nourishing all living forms and beings—all of her children. There again with her, we can partially sense and feel our, natural human part in this orginal play; before the edited, modern version began its run at the urban playhouse, the city.

'Civilization' socioculture has continuously slandered living nature's human nature, attributing bad human traits and phenomena to human nature, instead of to their two real origins: (a) 'civilization's adulterating artificialization of natural nurture and society and (b) episodic, incidental, anomalous features in natural, human society that do not endure for long in the play of human evolution. All our beauty and reality, well proven for so long in Earth-folk and primordial socioculture, we owe to our Mother Nature, the human nature part of her and our continuing honor of these two living natures along with evolved human nature's natural, nurtureculture, and society.

Our mother I, however, has a problem; she has been forced by historical force to mate with a force unworthly of union with her. After an eternity of monogamy and faithfulness with a mystical heavenly father—or an eternity of immaculate conception by a different metaphor—she now has a strange bed mate. She does not commit adultery; nor, as some would relate it, is she seduced. It is rape. How has this happened, the rape of the timeless, faithful wife of father sky—or with the other metaphor, the virgin mother of Earth-life! The loins of living nature have been violated by the only instrument 'evolved'—more precisely, emerged through transmutation—that is capable of doing so, the egoself's phallic, probing intellect. Is this, then, a self-violation, with the instrument being living nature's own natural, human intellect executing the human will?

On the contrary, the natural intellect and the transmuted, quasi intellect of 'civilization' are not the same; and, the twain shall never meet. The violating instrument is an alien entity: the transmuted, quasi intellect and egoself 'bonded' as a quasi and pseudo entity. The real and natural, human intellect has been faithfully held within the natural intergrowth and interfunction of natural mind and intelligence—the interplay of intellect, intuition, senses, instinct and emotion—from the start of the human family on through the genus, species and our own subspecies. Nowhere is it written on stone, spoken in lost folklore or predestined within the nature of living nature or within human nature, that human nature and human being are to self-destruct upon reaching our subspecies. Rather, something alien happened to living nature's human nature and being: they experienced alien transmutation through the penetrations of alien, artificial elements-phenomena of 'civilization'. The result was fragmentation of the natural, authentic, human mind, self and lifeworld–the fragmentation of these in themselves and from each other; all of which are intergrown, interfunctional and unified in natural, human community. These impacts repositioned intellect–now as quasi intellect–to the leadership of mind; and, it also split ego away from whole human self. These two merged together as intellect-egoself and commenced to misperceive–with tunnel vision and (in modern science) scopicized vision–and to study and organize (with the transmuted, quasi intellect) the fragmented ruins of whole self and lifeworld. These ruins were originally unified in the natural, organismic structure-function of human being, nature and reality.

Of course, the transmuted intellect-egoself cannot know it is a fallen version of the natural, whole self, i.e., not without a deep revelation into its remnant human being; a human revelation.

It is, then, 'civilization' humanity, as opposed to Earth-folk society that has been violating Mother Nature and her human nature, being and reality. The instrument has been the transmuted, intellect-egoself: the swollen, quasi intellect and transmuted, swollen egoself united as an alien entity. Alas, we are the aliens we search the sky and the imagination to find.

## 3. Human Environmentalism

Our challenge as environmentalists is to set about to do for human nature and being the same that we intend to do for living nature: to clean up adulterated socialculture and preserve the natural, human nurtureculture and nurturome of our rightful, human selflifeworld. We cannot limit our cleansing and protection to material, geographical matter; but, rather extend this to our polluted, deteriorating persons, families and communities of sci-technic sociocultures. We must recognize and remove the adulterants that pollute our minds, relationships, being, spirit, reality, perceptions, feelings, thoughts, work, play, purposes, and our hearts and souls.

What a huge, conspicuous presumption it is that while environment–living nature with all her ecological species including our own bodies, minds and health–are being infected, damaged and deteriorated by the pollution of sci-technic, 'civilization' socioculture; that yet, in spite of this happening, by some, mysterious immunity, our own beauty, goodness, reality, our own consciousness, being, thoughts and feelings are presumed able to escape from this all pervasive deterioration through adulteration. These nonmaterial elements of our life and being, these also have there *natural* conditions and 'environments' for growth, fulfillment and perpetuation! Should we not, at last, become the protectors of the environments of human being's immaterial realms, growths, and their harvests.

After four billion years of the ongoing triumph of evolutionary creation-perpetuation of life over the dead star-stuff, the universe, modern science-technology arises to tell us that it knows how to save our living planet and humanity from apocalyptic decimation. This is the arrogance and stupidity of the over educated or miseducated. Which of these two shall we trust? The contemporary science-technology that is married to the very power-wealth ethos that has been for 300 years destroying the planet? Or, living nature and her ecological laws and wisdom which have won now, for about four billion years running, any Good Earthkeeping Seal of Approval imaginable? Living nature, has, certainly, earned a Seal of Approval from humanity, and got it–prior to Western domination and violation of her, i.e., in the form of ceremonies of gratitude and reverence to a sacred Magna Mater and, later, a Gaia. In truth, the Earth approves herself, will approve of and perpetuate us if we approve and obey her natural and ecological laws of life, for which we were also evolved, ecologically to be within; laws we incrementally violate in our unknowing self-violation and in our violation of her, reaping returns not fully imaginable.

# 4. Truth and Reality

One basic, modern problem inherent with sci-technic 'civilization' is that truth and reality become separated into a disharmony: what is regarded true and what is regarded as real are no longer synonymous. The Western intellect has expanded apprehended objects and phenomena beyond what the natural mind and consciousness authentic to our human being, nature and identity normally and correctly apprehend. This expansion of apprehended quasi reality envelops alien elements-phenomena into our natural, authentic human reality; thereby not just adulterating reality with unreality, but human truth authentic to our species with quasi truths–falsehoods that do not belong to/with, apply or fit into authentic human reality. An example of alien knowledge is telescopic knowledge of the moon, solar system, galaxy and universe: knowledge acquired from means outside natural, human faculty–outside natural senses, instinct, intellect and intuition in their natural, balanced interplay. Telescopic vision, as well as microscopic vision, is outside the natural function of the human eye; it affixes alien lens to the eye.

"The proper study of man is man" is only one third of the truism: the proper study of man is his natural human being (and beingness), and the natural human reality and human nature indwelling to this natural human being. The truths or knowledge of this natural being and reality is of them, apprehensions of human nature, being and reality. Truth–natural, authentic, human truth–is grounded in natural, authentic human reality and being; and is revelation or disclosure of elements-phenomena (truths) of these to human being's natural, authentic consciousness and self-consciousness. When consciousness and self-consciousness are adulterated and disintegrated, truth and reality are separated and disunified as both become reflections of blended, quasi being and reality.(*See* scopicized understanding in Ch. V.)

The main objection that some will raise against the above account of truth, reality and natural human being, as well as, against the account of natural, authentic human socioculture given, herein, throughout, is the common assertion and belief that humankind could not have survived and, further, cannot survive without science and 'civilization'. This belief is adequately discredited in "Natural, Authentic, Human Society vs. 'Civilization'" and elsewhere in this volume. Briefly, here it will suffice to mention three counterpoints. First, the discovery by cultural anthropology that the Earthfolk sociocultures of the North Atlantic European fringe escaped generally and essentially the reaches of Western 'civilization' from the hunter-gatherer stage through 3,000 to 8,000 years of agricultural Earth-folk, a'civilization', village lifeway to approximately the start of the 20th century. Second, it cannot be tentatively held that after several millennia of agricultural self-sufficiency and over 45,000 years of hunter-gatherer self-sufficiency that these humans of our modern species need the adulterous penetrations of 'civilization' nor of 20th century 'civilization' for survival and perpetuation. Third, at least hundreds of millions at the present, live in self-sufficient, agricultural, Earth-folk communities, though in rapid decline under the siege by 20th century science and technology, supposedly our new agents of salvation, consciously and actively resist the threatening, alien, adulteration of 'civilization' lifeway that, if unhalted, replaces family and community self-sufficiency, cohesion and autonomy with poverty, fragmentation and dependence.

For 5,000 years educated man's inquiries, debates, confusions, questionings, studies, speculations, doctrines, systems and philosophies have stemmed from the mistaken perception and conception: that he and the 'civilization' man he represents is representative of the predominant and general human condition; that the human condition entails 'civilization'; that modern humanity, our subspecies, now dated back 50,000 years, is inseparable from the artifice of 'civilization'. The revelation of human authenticism frees the intellect from the false foundation or core perception and conception upon which or around which, respectively, the above products of the intellect are constructed; it frees the intellect from its exile-prison to rejoin and reunite with the natural, authentic whole mind–

to interplay, interfunction and intergrow with its family—with intuition, instinct, and senses. It frees the intellect, as slave laborer to fallen, inadequate egoself, to rejoin the community of whole triecologic self—bio-self, family-self and community-self. Together again, these will intergrow with reintergrown, natural lifeworld and its natural environment. Back again into our structure-function-dwelling, selflifeworld, we can with family-community-village recover from alienism, regrowing human authenticism our authentic humannness.

# 5. Metaphysics and Epistemology

Human authenticism reunifies metaphysics and epistemology by revealing: that reality is not a matter of real world but of real selflifeworld; that there is no such thing as world *per se* by itself, without life, because world entails a living entity/organism to sense or experience or have indwelling being within it. Inanimate matter, whether a grain of sand, a planet or galaxy, is not and does not become part of world—lifeworld—until experienced or interfunctioned with an organism. Inanimate matter is to life as corpse is to living entity: the inanimate is without world and life—without lifeworld interjoined, interfunctioned—because life lives, i.e., exists, both as structure-function (bio-organism) and organismic structure-function-dwelling (bio-social-spiritual-ontological, organismic being).

Reality exists only of, in the context of, a living entity: each kind—species—having its reality—its real world, life and self—its intergrown, interfunctional selflifeworld. An inanimate particle, planet or galaxy has no world, nor is a world, until incorporated into a living, organismic selflifeworld—they remain until then objects: world objects, not lifeworld objects; moreover, they are alien (quasiworld) objects when viewed through alien lens. This difference between sphere-world and lifeworld must be realized by humankind because the confusion, along with egoself's other key confusions, is literally destroying our species and its future.

To illustrate, planets percieved by the natural eye and mind are lights in the firmament; but perceived through artificial, alien vision—

telescope–the perception of them is transformed into the perception of large material objects which entails the study of them, a quantum leap from occasional, natural, human sight of them. Natural, human sight and perception are parts of natural, human mind, being and their natural reality; alien sight and its entailing, alien perception, mind, consciousness, being and reality are outside our natural sight, perception, mind, reality et al. Planets have two existences: their own apart from our perception of them; and their existence as firmament lights–lifeworld objects–percieved by a natural being's natural sight and perception. Their existence percieved by alien sight and perception is outside human reality, mind and being. Any alien sense–whether alien and artificial, e.g., telescopic sight or merely alien, e.g., dog eyes transplanted into a human–cannot fit into or participate in the natural perception, mind, reality and being of a natural organism.

The sensing (e.g. sight), perception, consciousness and thought through an artificial transformation of sense, e.g., through telescope and microscope, are all alien, artificial phenomena outside of human reality, being, spirit and nature; and cannot participate in these or there selflifeworld. The telescopist and microscopist scientists are deep into hallucinogenic mind, consciousness and unreality; deeper even than the regular users of hallucinogenic plant chemicals which at least are natural to living nature, and which at least some animals ingest sometimes purposefully.

There is a line between the science or philosophy of living nature and human nature, and the science or philosophy of the artificial: the first is a humanism and has at least some chance to manage and solve the natural and alien problems of 'civilization'. The second is alienism and is the manifest, progressive movement into the artificial–artificial things and artificial life; and artificial life is a contradiction: life is living nature, especially as species being.

Reality, then, of an object or phenomenon exists only of or in the context of their having meaning which ultimately means that they must be related in some way to the purpose of a living species or species' selflifeworld(s), or the purpose of Earth's entire ecological system which is to support and interbalance all living species' purposes.

By revealing that our human subspecies–like all species–has a selflifeworld that is fitting, correct and true to us; that is good, right and real for us; that is utterly authentic and natural to us; human authenticism unifies all branches of philosophy into one body and vision. Our truth, reality, mind, past, goodness and beauty are revealed to be nothing other than our natural, authentic, human nature, being, identity, reality, and spirit within their nurtureculture and selflifeworld within Earthlifeworld (ecosphere).

## 6. *The Cosmos* of Carl Sagan vs. Earthlifeworld

(Written in 1993 before Sagan's death in 1996.)

Carl Sagan asserts that the exploration and colonization of outer space—our solar system and galaxy—will be more exciting, enjoyable and fulfilling than our previous age-old exploration and colonization of the Earth's regions, namely, the exploration and habitation of wilderness areas that humans have undertaken from prehistoric human migrations to modern age fronterism. Sagan and his followers don't explain why and how this can be true. They don't give the philosophical or rational basis for this claim that space exploration and colonization of inanimate—or almost entirely inanimate—planets and moons, having few of the essentials for human or plant and animal life, such as air and water, should logically, psychologically or philosophically be more enjoyable and fulfilling and adventurous to humans. Nor do they explain how this can be anything else but an extremely difficult, costly and alien challenge. Afterall, the exploration and human settlement of the Earth is directed by our genetic nature given to us from millions of years of evolutionary growth and our more recent human species nature given to us through natural evolution as natural beings within the natural, ecological system of our Earth.

Carl Sagan says the urge to go, to see, to know is seen throughout human history and characterizes the human species. The truth

is that up until the age of discovery, only a few individuals experienced this urge (.001 % of humanity?). And, furthermore, even
with the explorers and pioneers of about 1500 to 1900 A.D., it
was, instead, the desire of a few to be wealthy and/or powerful
that, along with the will of more than a few toward a better life in
frontier lands that propelled the movements and migrations to the
New World. The truth is that even the many migrations that have
occurred throughout prehistoric and historic times, were migrations always on the part of merely a small minority of each generation of humanity. Further, they were spurred by drought, disease,
conflict, competition for game or land, and other phenomena separate from "the urge to go, to see, to know".

In one of the 1990 "updates" in *The Cosmos*, Sagan, apparently accommodating or being influenced by an increased environmental awareness states: "the Earth is the only planet in our solar
system that contains life and it is our responsibility to cherish it."
But, cherish entails not just a feeling in the heart but protective action. And action in a system of finite resources and endeavors, entails
priorities. Whenever one of the essentials of collective survival, e.g.,
the environment, becomes threatened, it becomes the top priority to
secure that essential. Top prioritization, inturn, entails the sacrifice of
any other needs and wants that are not essential. Space exploration
proponents respond with the contrived assertion that human nature
harbors a psychic need to explore space, a need that if not met will in
some obscure way mean decline for humanity.

This debate, then, over whether or not to explore and colonize
space hinges, not just in large part but entirely, on whether or not
this activity is within the genuine essential needs of human nature
and being, of human spirit and reality. The answer is unequivocal.
We humans were created either out of the dust of the Earth (the
Biblical account) or out of Earth's evolutionary life (the scientific
account). Our human nature and being, therefore, were created
either from Earthly matter by a God's intelligence or from the
evolutionary activity of the Earth, with or without God's direction. It logically follows from our evolutionally created living na-

ture that Human needs exist, solely from our present human nature and any past stages of humanity's evolved living nature and
creation. The only genuine, human need to explore the firmament
is limited to seeing, thinking, and wondering about the night sky
and the sun of day. This is easy to understand, 'as clear as the sun
in the sky', except to those propagandized by contemporary astronomy and the sciences associated with it in vested, special, interests. Saganite scientists confuse (a) the natural wonder one feels
upon viewing the firmament through natural sight, with (b) the
career need they feel when viewing the sky through the telescope.

The great misfortune and error that befell Carl Sagan and
many others that place, in importance, the vast outer cosmos
before human life and Earthlifeworld is that they began, at an
early age, perhaps six to eleven years old, to discover and be
captured by the wonders of the solar system and alien outer
space before they were able to discover and be enraptured by
the wonders, fascinations, joys and blessings of their own natural
world through direct use of their own natural senses and intelligence. Television, books and the classroom got to them before
they could get out to their real source and connection, living
nature.

Sagan believes "Our voyage of self-discovery" takes place "on
our journey to the stars". This is preposterous: to hold that each
generation of a natural species is not evolved, ecological by living
nature with the capacity to discover itself right on its own terrain,
in its own biosphere and ecosystem from which it was evolutionarily created.

The other basically, nonsensical notions and beliefs of Saganism
need to be enumerated and refuted for the sake of human wisdom
and survival. A counterpointing of *The Cosmos*, Carl Sagan, 1990,
is called for, herein. (Direct quotes and paraphrases are from the
1990 revised edition.):

(1) "We are star-stuff that has taken its own destiny in its own hands". No, Mr. Sagan; we are Earth-stuff: about four billion years ago a new-stuff was created—Earth-stuff. The Earth became distinct from inanimate star-stuff (notwithstanding insignificant cellular life on Mars) with the creation of life and its living evolution. In truth, we are living stuff, participating in the unique, vibrant movement and drama of evolving life and its (life) world, the biosphere (Earthlifeworld). The failure to embrace and identify with life and its ecosystems: to mistakenly identify centrally—or even equally—with inanimate matter is the great tragic failure of the millions of lay and professional Saganites. Amazingly, they have come to a philosophy of life that centralized non-life—or at best equalizes the two.

(2) With Saganism, the highest value is placed on those objects and activities that eventually "will help to carry us to the stars." According to Sagan and Saganism science, we are on a journey from the stars, from which our star, the sun, was formed. However, it is much more relevant to us that after several billion years, our planet was formed and upon it about 4 billion years ago, there emerged a second type of evolution, namely the evolution of life, an evolution distinctly, dramatically, and wonderfully different from the evolution of inanimate matter. And, that from this living, vibrant evolution, we emerged, Homo sapiens, after three or four million years of ape to human evolution. Why, then, should we, as the Saganites propose, commence a journey back to the stars, to inanimate matter from which only the *lifeless* Earth evolved—not us! Afterall, Sagan admits that Earth's evolution of life is distinct from any life that may exist beyond our reach. Why should we seek to leap backward past four billion years of our kind of evolutionary creation—evolving life—to dwell amidst and colonize nonliving, inanimate, alien, cosmic matter! Why should we commence now a journey back to the stars, and star-stuff, to inanimate matter! Why not, instead, continue our journey forward into the living evolution of Earthlifeworld and embrace *our* evolution, of our living, human being-stuff!

(3) Our origins as living Earth beings lie amidst the Earth's biosphere, amidst Earthlifeworld, and our destiny can only be to continue that journey of Earth's natural human evolution: to preserve our natural species and its natural elements and ecological essentials against the alienization and artificialization that forebode its extinction or decimation.

(4) Why should we chose to embrace the lifeless evolution of dead matter over—or even equal to—the four billion year old evolution of life that created our own 300,000 year old human species! Why seek human destruction outward into the alien realm instead of continuing onward with the cyclical growth-fulfillment-perpetuation of Earth's evolutionary creation-perpetuation.

(5) "The cerebral cortex is the distinction of our species". This, even though Sagan shortly before this statement, says that it originated with the primates. "Many other animals have feelings; what distinguishes our species is thought. The cerebral cortex is, in a way, a liberation. We no longer need be trapped in the behavior patterns of lizards and baboons, territoriality and aggression, dominance, and hierarchies." Sagan, here, conveniently forgets human history—'civilization'—modern ethology and contemporary human events and behavior that exhibit these supposedly reptilian patterns of behavior. "No longer at the mercy of the reptile brain we can change ourselves." In fact, the mammals and the primates, too, are no longer "at the mercy of the reptile brain;" but, we don't see them changing themselves via adulteration, nor sitting on stools looking through telescopes and microscopes in a deep search for who and what they are. The elephants, dogs, primates and many humans still know "who" and what they are; for they are living it: they live their own, true, real, authentic, natural being, each created by living nature to fit into and have a place in the magnificent, wondrous, triumphant movement of lifebeing-force within Earthlifeworld.

(6) "We are each of us largely responsible for what gets put into our brains, for what as adults we wind up caring for and knowing about." Here, Sagan conveniently sweeps aside accepted ideas in infant and family psychology and in sociology concerning the formation of personality and thought and feeling patterns.

(7) "We are the only species on the planet, so far as we know, that has invented a communal memory, the warehouse of that memory is called, the library. The evolution of human intelligence: from genes to brains to books." Not merely forgetting that some intelligent 'civilizations' had no written words or books, such as the earliest period of the Mayans, Sagan puts down and insults the intelligence of pre'civilizaiton' humans; and, further, the intelligence of 'uneducated' rural peoples of our time and throughout historic times. Sagan, like most educated people, doesn't know what intelligence is—or forgets. The equation of intelligence with formal education or book knowledge is a common ploy and blunder. He forgets or doesn't know about the existence of the communal memory before books: in the form of folklore, dance, ceremony, customs and mythology. This communal memory still works today for billions of uneducated—but equally intelligent—people.

(8) Commenting on the Voyager project undertaken some years earlier, Sagan remarks, "They will know from this Voyager sent into the cosmos that we are creatures with at least a little intelligence and a longing to make contact with the cosmos." However, he speaks here only for Saganites, scientists and layman alike. In truth, this 'longing' to make contact is felt by less than one tenth of one percent of about five billion human beings. Sagan's words underscore the grab for power and usurpation of authority especially typical of Saganism and typical of science in general. Democracy is flagrantly ignored: the will of the masses becomes something to be molded through media powered propaganda from whatever power-wealth, special-interest group that can pay for it (in this case, a school of the astrophysical sector).

(9) "In a dee*per se*nse the search for extra-terrestrial intelli-gence is a search for who we are." Actually, depth or profound understanding means basic or essential truths about the nature and reality of something—in this case the nature and reality of human being. Since Sagan states elsewhere in *The Cosmos*, that any intelligent, extra-terrestrial beings we may have contact with will not be much like us—will not be human or humanoid, here, he is, therefore, both (a) flatly contradicting himself and (b) speak-ing nonsense about how to understand who we are. The search for a profounder understanding of humans actually must be centered upon four subjects: contemporary humans, our historic and pre-historic ancestors, our predecessor human species—Homo erectus and Homo habilus, and, circling outward to primates and social mammals. The search never goes outside the process of evolution-ary life that created us, never into some discovered, alien process of creation and its intelligent creatures (unreachable for study). Such a study would only reveal knowledge of who and what we are *not*, which we can already know sufficiently by (among other ways) reflecting upon transmuted, adulterated human beings in high-techno society, one type of which is exemplified by the Saganites amongst us.

(10) "Every culture on the planet has derived its own response to the riddles posed by the universe. Every culture celebrates the cycles of life and living nature." But which universe? The universe of pre-telescopic cultures, consciousness and perception is a differ-ent one from the universe, the cosmos of which Sagan speaks and studies. Referring to nature, Saganites include outer space into nature; while, pre-telescopic societies of natural perception, con-sciousness and intelligence do not: they speak of the visible, natu-rally perceptible Earth and sky as nature.

(11) "The diversity of the phenomena of nature is so great and the treasures hidden in the heavens so rich, precisely in order that the human mind shall never be lacking for fresh nourishment." Here, Sagan, like a quick, smooth magician, pulls the very distant, unreachable, nearly irrelevant outer space down to Earth, grouping outer space with the Earthlifeworld that envelops, creates, sustains and immediately involves us—our past, present and future—within her rich, living organismicity of Earthlifeworld. In truth, Sagan's perception is distorted: It is rather the diverse treasures of the living Earth that we are involved with, that have and will provide "fresh nourishment for the human mind" and for human being, spirit and nature as well.

(12) In speaking about science, Sagan says that it is another way to express our longing to be in touch with our origins; and, that it too has its rituals and commandments. In speaking of science, Sagan attempts to affirm its humanness, its relevance and inherent link to humanity. He fails on two basic counts: (a) "Its only sacred truth is that there are no sacred truths." However, every human society has held some truths, or truth—and some reality—to be sacred or ultimate. Since truth and reality are intertwined, Saganism holds that there is no human reality except the never-ending search to understand and find ourselves—we are lost forever in an infinite search since "there are no truths"—and, therefore, no reality—about ourselves. None, that is, except for one, which Sagan dares not state clearly: that now and in the future 'humans' do and will create themselves in new worlds (new planets) with new sciences such as genetic engineering. Nothing is sacred—inimitably true, good and real—except the ongoing re-creations of a quasi humankind and its quasi lifeworlds, creations performed by elite cores of specialistic scientists. Should human beings laugh or cry or feel pity at this breed of scientists? Or, rather, should we fight them while we still can!

And, again Sagan's attempt to humanize his breed of science fails with (b) "with this tool we vanquish the impossible." However, every species has boundaries to its activities and functions, to its life and being. The 'humanity' in a Saganite future has no limitations or boundaries because their quasi nature and reality is re-creation itself, i.e., experimentation, testing, redesign, and re-creation in perpetuity. Surviving to experience this exhilarating thrill of power is their ethic and reality here. Theirs is an addicted breed of quasi humanity: they exist for the rush that comes from scientific experimentation and re-creation of quasi humanity, falsely based and doomed to failure and decimation or extinction.

(13) "We've begun at last to wonder about our origins, star stuff, contemplating the stars. . . . contemplating the evolution of matter, tracing that long path by which it arrived at consciousness here on the planet Earth and perhaps throughout the cosmos. Our loyalties are to the species and the planet. We speak for the planet; our obligation to survive and flourish is owed not just to ourselves but also to that cosmos, ancient and vast, from which we sprang." Sagan, right after saying that "our loyalties are to the species and the planet," immediately grants the cosmos an equal loyalty to the Earth: "our obligation to survive and flourish is owed not just to ourselves, but also to the cosmos. . . . from which we sprang." In truth, we sprang—evolved from earlier life forms that evolved from the very first life forms created on Earth—not in outer space—and all were nourished by Earth stuff—not outer space stuff. Life forms that, moreover, Sagan admits are unique to Earth and distinctly different in nature from any other evolutionary life systems that "perhaps" exist somewhere in (practically speaking) unreachable outer space. Our loyalty and obligation to—and our emotional and spiritual attachment to—the cosmos is correctly minuscule beside our relationship to our nurturing mother Earth.

About four billion years and the wonderful phenomenon of living evolution—our nurturing, originating evolution—powerfully necessitate and make obvious the distinction between the living Earth (including us) and the dead cosmos we behold (with artificial vision) so far away. And yet, Saganites, with their vested, career interests and misdirected childhood fascination, cannot and will not see the obvious important difference. Until Sagan and the Saganites correct their pathetic and tragic distortion of priorities, their can be no forgiveness for them.

(14) In Sagan's 1990 update to *The Cosmos*, he points out "that some of *The Cosmos's* boldest dreams about this world are coming closer to reality. Since the broadcasting and publication of *The Cosmos* in 1980. . . . the imperative to cherish the Earth and to protect the global environment that sustains all of us has become widely accepted. Perhaps, we have, after all, decided to choose life." A strange, tricky contradiction to the foremost purpose of *The Cosmos*, i.e., to explore and understand the lifeless star stuff beyond the Earth and outer space.

(15) "It is as if there were a God who said to us 'I set before you two ways: you can use your technology to destroy yourselves or to carry you to the planets or the stars. Its up to you.'" Here, Sagan reaffirms the two beliefs that provide for human extinction or decimation via the unhuman alienism of Saganism One, Sagan, despite saying moments earlier that our first loyalty is to the species and the planet, contradicts himself and reaffirms that implicitly the highest goal of science-technology is to "carry us to the planets and the stars."

Two, by affirming, in effect, the traditional Judeo/Christian doctrine that man has created with both good and evil within him—in this case with the capacity to use science-technology for our benefit or our self-destruction—Sagan is reaffirming the historical effects of that doctrine: namely 5,000 years of cyclical rise and fall of Western, 'civilization' sociocultures. But, this reaffirmation about human nature and technology has recently become a gamble, with the stakes being humankind's survival or extinction, because the next self-destruction forebodes as the final one with no following reemergence of 'civilization' or even the human species. These two doctrines about human nature and science-technology (especially in the 20th century) have been rendered both untenable and suicidal by the progressive destructivity to humankind and the Earth (environment) of post war sci-technic socioculture.

(16) In closing, Sagan's apparent life mission and the mission of *The Cosmos*, to advocate the exploration of the solar system and our galaxy, collided in the late 1980s with an ongoing body of Earthly understanding and caring: a caring for the Earth as our only fitting, hospitable home within any appropriate human consciousness—a home threatened with environmental, nuclear and other forms of destruction. About ten years after *The Cosmos* was published Sagan came to declare that our first loyalty is to the species and the planet and that "it is our responsibility to cherish it." But how much of our available devotion and loving efforts—after allotment for ourselves, families, communities and societies—should we give our species and planet? Is 50% enough, as implied in his statement "We are equally children of the Earth and sky." (by which he means Earth and outer space)? Giving outer space 50% of our devotion, efforts and resources doesn't make sense with the Earth now threatened with destruction. "We are taking some initiating steps; but, again, these steps are too small and too slow." Sagan cannot see that it is one of his own doctrines that slows and perhaps dooms to failure our rescue of the Earth: the doctrine that we are equally children of the Earth and outer space.

## 7. Beauty

Something may be beautiful, have elements of beauty, without being beauty itself. Beauty itself is that which has its natural parts and functions without adulterations, freely fulfilling natural purpose and system: fulfilling natural law whether of living nature or of human nature and being. Man made 'beauty' is beautiful to the degree that it imitates or portrays the simple purity of natural ecosystem. A person is beauty itself when action or mental activity is purely constituted of elements of our natural being. Real art strives to imitate beauty itself by knowing and utilizing the elements of natural ecosystem, and is beautiful to its degree of success. Art can never be beauty itself, but is a purgation of some of the superfluities, the adulterations introduced by 'civilization' into the natural ecosystem and natural human ecosystem.

In truth, it is only some features or elements of beauty that are in the eye of the beholder. There is, in fact, available to human perception and awareness, both subjective and objective beauties. Indeed, there is an underlying objective beauty recognizable to us just as there are underlying, objective, human truths applicable to all of our greatly diverse humankind. Beauty, culture and personality, in one respect are exactly like truth, culture and personality: beneath the subjective elements of culture and personality, there are objective, human truths and beauties.

Further, since few would deny that human reality contains some human beauty, or has some beauty within it, the pursuit of, the definition or description of reality includes the pursuit of the definition or description of beauty. To acknowledge that there is objective truth and reality within the realm of our human species that is, through our species evolved system, fitting and good for us to know and be conscious of, is to acknowledge the same about objective beauty.

It is the philosopher's task to recognize and teach of these objective elements-phenomena of human reality. It is the artist's task to recognize them for the sake of their nourishing and healing effects upon human being.

# 8. Earth-life-Sun Cosmos vs. Universe/Cosmos

In the modern era a struggle emerged that has now become a great ideological and philosophical struggle. Upon the outcome hinges the fate and welfare of humanity and of the Earth—the magnificent home of all life (we know of, or is reachable and relatable). It is the struggle of two ideas for the heart and mind of humankind: the Earth/sun cosmos vs. the infinite universe cosmos. The first idea, the older by far, is that the Earth/sun cosmos is the world of humanity; the realm we humans deal with directly and continuously in thoughts, activities and feelings. This is the world we can know very much of and have experience with. The second idea is that outer space, the universe beyond the Earth's atmosphere and beyond its sun and sunlight itself should be our main realm of reference, or should be humanities newly discovered second realm of reference and study, either equally deserving our attention or even a major part of it. This second realm, outer space, is the world that is infinite, largely unknowable and entirely unexperiencable by humankind, except via two opposing ways: (1) through our natural vision and consciousness of the night sky; and (2) illegitimately, unnaturally and unfittingly through the telescope, television and other artificial instruments.

To use the telescope and other artificial instruments to see and contemplate the firmament is to miss seeing, perceiving and experiencing the firmament as a natural being does—as a human being does. This removes the *human* vision and experience of the firmament and the human imagination and wonder about the firmament from the human spirit and human being. How many astronomers, amateur or professional, spend some time as did pretelescopic humans, sitting outside after nightfall experiencing the firmament in ways it has been experienced within the consciousness of awe and wonder of our natural, human being! Modern astronomers, instead, contemplate the nature, configurations and arrangements of inanimate matter from the perspective of mathematics and the physical sciences. They also seek and long to make contact with technologically advanced, alien forms of life. Such fools they are; for, they have become alien to the Earth's living nature and her human nature and being. They have partially become the aliens they dream of finding.

## 9. Astronomy, Outer Space, and Living Nature

Scientists are trying to illegitimately and falsely incorporate outer space and alien realm into our realm, our scheme of things, our system of Earthlifeworld and its ecological, human selflifeworld. If they succeed, it will be a rubiconic mistake and lie, concerning the nature of human being and spirit. It will be another key illusion and lie about ourselves and our selflifeworld, and another adulteration, diminishment and fragmentation of living nature, our human nature, spirit, being, reality, and the consciousness and understanding of living nature and its natural, authentic, real, human life (selflifeworld).

Our mistake is in believing that 'God' created the universe when in fact the universe created any 'God' we may have spiritual experience or relatings with, the only 'God' we will ever have which is our mother Earth, or her Spirit or life force which creates evolutionally human spirit and any other spirits one may wish to acknowledge of animals and plants. Earth is the only source of life, spirit, and being we have found, have evolved from, will ever know and be able to make contact with and participate in. *Our proven* 'God' is the body of Earthlifeworld (ecosphere), the Earthlife force and the spirit of all that life and its myriad forms created itself, evolves itself, came into evolutionary existence. Now, astronomical science has pulled a switch on us, pulled a trick, by conceptually transmuting the very nature of our 'God' through extending our knowledge of the alien portion of creation—the outer universe—and beguiling and persuading scientific humanity to believe that our 'God' created all of that new universe we are continually dis-covering and, presumably, will continue discovering into our infi-nite future. Science in this way lays the conceptual groundwork to maneuver its way into the position as the highest priesthood for humanity. Since we perceive and worship any 'God' through what we believe that 'God' has created, astronomic and microscopic science are now in the position to claim that all further knowledge of the nature of the creation—and any creator or 'God' of that creation–all such further knowledge will be delivered by science. Science in this way can replace the prophet, saint and/or holy man as the receiver of divine revelation with the scientist as discoverer of pseudo-mystic, secular, cosmic, alien revelation. In this scheme alien, cosmic revelation thereby replaces spiritual-ontological rev-elation. New data on the inanimate and/or imaginings about pos-sible or probable lifeworlds beyond our conceivable ability to reach or make significant contact with: this replaces any intuitive or tran-scendental revelations or experiences we may have and embrace about the nature of our animate, human spirit, being, nature and reality, and its relations to other living species of beings and spirits (of animals) and any source of unifying Spirit and Being underlying or existing throughout a knowable and experiencable Earthlifeworld—the only lifeworld we can ever foreseeable hope to know and experience.

Scientists yearn and fain to take spirituality out of the evolved, organic, spiritual-ontological realm of experience and root it within the world of the theoretical imagination—the scientific, hypothesized, theorized imagination. That is now their most shameful fault, the attempt to replace spiritual experience with the excitement for new data, the fascination for 'possibilities' and the adventure of discovery *per se*: to replace the richest human *experiences* grounded in our natural, magnificent, human being and natural lifeworld with *imaginings* grounded in the artificial and the inanimate matter of unreachable, insentient outer space.

Astronomy's idea that we can get closer to living nature or know more about living nature through the use of artificial tools or technology, such as the telescope and the radar scope, is utter nonsense and even sadly pathetic—in fact, a pathetic contradiction. To realize this, the nature of living nature must be realized and defined, i.e., that it exists as two basic realms. First, there is the nature of our own natural, human subspecies, e.g., its consciousness, being, reality and spirit: this is the realm, human nature. This first nature is a part of the larger realm of living nature, i.e., the realm of all living things and the inanimate matter which participates in this larger living realm, which we have historically and traditionally termed nature in general. A third realm is outside of and alien to general living nature and its human nature realm. This alien realm consists of three basic parts. First, there is all that beyond the Earth, moon and sun (and beyond our solar system) which cannot and does not participate in the natural, human consciousness authentic to human nature and being beyond its merely being beheld by natural sight as the firmament and perceived by natural, human mind and consciousness as the firmament. Second, there is all the rest that modern science perceives, including telescopic and microscopic sight and studies. And, third, there is all else that is outside naturally perceived and known living nature and her human nature and being, i.e., all artificials created by humans, all technology and activities which we humans can only behold and engage through utilizing artificial, alien objects and/or phenomena such as artificial invention, science in general and artificial, social activities, organizations, institutions and units.

Astronomy studies all that which is beyond our natural sense of sight—that which is very distant and unreachable from us. While science in general studies all else which is beyond the natural senses, natural perception and consciousness (including microscopic objects-phenomena, i.e., all that is outside of and incapable of participating in *natural,* human consciousness, mind, being, life, reality, spirit et al.; all of which to reality is alien—outside of living nature and human nature; but which is now misperceived by scientists to be part of living nature through the fact that their (transmuted) intellects (i.e., bio-intellect-egoselves) are capable of beholding and studying it with transmuted, alien 'intelligence', intellect, and alien, artificial device. Tragically, they are unaware that a mutation—a new kind of mutation, alien mutation—has occurred which produced the intellect and the bio-egoself which joins it, creating the mutant, alien bio-intellect-egoself which though interlinked to natural, biologic body, cannot function as nor be restructured as natural entity, from its being constituted of partial, alien, artificial structure-function.

In other words, and from another direction, should the word *nature* and the phenomenon it indicates be applied to what we humans naturally and normally perceive and are conscious of as natural to our realm and to our consciousness? Or instead, should it be extended to all the universe which we are still discovering and will discover in the future through the use of *artificial* vision, technology such as the telescope, and adulterated, transmuted perception? The answer is certain and easily determined. What we can through general consensus regard as 'natural', using natural sight, the other senses and natural perception and intelligence is what constitutes living nature. How can something be a part of living nature if it *cannot* possibly be perceived by some natural perception or come into (enter) the natural consciousness of some natural being (animal or human animal), i.e., enter some consciousness evolved by an evolutionary system and realm of life. Nature, to have meaning and purpose entails living animal consciousness.

Outer space has no naturally occurring—springing from natural consciousness—meaning except what it means to the natural eye and consciousness. It is outside natural reality and natural realm which entails some natural consciousness. Further, human reality entails purpose within natural consciousness. Alien purpose occurs in alien consciousness of alien being and realm. Like the alien, artificial consciousness of astronomers and astronomy, respectively.

All of creation, then, is not living nature; since, the consensus has been and still is that whatever intelligent life forms that may exist, and/or may visit us from other solar systems and galaxies, are to be regarded as aliens. Their worlds must also, therefore, be alien: *alien to living nature*, to Earthlifeworld. Earthlifeworld being natural to itself, its life and us. When contemplating worlds or lifeworlds with life forms, there then are two opposite and mutually exclusive realms: natural realm and alien realm. Merely because they were, (in accepted theory) some several billion years ago, the same realm of exploding, inanimate matter, and ions later are still part of the creation (universe), astronomers would like to include this infinity of exploding, inanimate, infinite matter with living nature. However, most of the universe does not participate in the natural realm of human selflifeworld nor in the natural realm of Earthlifeworld, except illegitimately—when forced into our perception by astromomers: a violation of our natural perception and consciousness which is right for us, fitting and authentic to us.

Living nature, then, is (a) that which we can in theory be naturally conscious of through natural sight, and/or senses, and natural perception and (b) that which participates in theory or practice in some small, significant way in natural Earthlifeworld (biosphere) living nature or in natural, human selflifeworld or within natural, human consciousness. The moon, sun and its planets qualify through meeting both criteria; the firmament beheld by natural sight and perception qualify through both criteria. Telescopic outer space meets neither criteria and is of alien realm, not natural realm.

Paradoxically, that which is observed through the microscope, from its primordial participation in Earthlifeworld, is of the natural realm; but, the beholding of it is not natural: we leave the natural realm including its human nature, being and consciousness when we look and perceive through the alien microscope (and telescope) or contemplate phenomena beheld through it: we partially become the aliens we scan the universe for.

# 10. Natural Perception

The purpose of the natural eye is to provide natural images of the natural lifeworld and natural reality for the natural mind to integrate into natural consciousness through the natural structure-function of the eye itself. When the natural structure-function of the eye beholds unnatural objects-phenomena, that appearance unto the eye constitutes a violation of the natural purpose of the eye which is to behold natural objects-phenomena. So, the mere beholding of and appearance by artificial objects-phenomena aside from further engagement of these by other senses such as touch, smell and taste, and engagement into natural mind, consciousness and activities of the natural organism which is utilizing the natural structure-function of the eye toward its natural purposes: this mere appearance before the eye (the beholding) is the first penetration of alien, artificial element-phenomenon into natural, authentic, human selflifeworld. The definition, then, of natural authentic element-phenomenon is that which, first, can be beheld by or made contact with (sensed by) the natural senses; and, second, which then can be incorporated into the natural structure-function-dwelling—selflifeworld—authentic to the natural species in question. That is to say, it can be incorporated and assimilated by the natural structure-function of the mind, i.e., senses, instinct, intuition and intellect functioning in their natural structure-function which is the incorporation and assimilation of objects-phenomena into the natural, authentic structure-function-dwelling of human being, nature, reality and spirit—the natural, authentic selflifeworld of the species or subspecies in question.

## 11. Human Organismicity

It is a law of natural, authentic human socioculture, as well as, a law of natural, authentic human nature, being reality and spirit, that the growth into and through social units and through sociocultures is timed and sequential. This is so precisely because human growth—biologic, social, sociocultural and spiritual—is, like all living growth, organismic. Non-organic, material, mechanical growth is actually material development by design of the artificial intellect (of bio-intellect-egoself) rather than by natural whole intelligence. If they are to be healthy, natural and harmonious, rather than decayous and sickly, human socioculture's central goal, purpose and policy is to adhere as far as possible to a crucial law of human sociality and socioculture authentic to our subspecies and species, the law of human organismicity: by natural law human growth, biologic, social and sociocultural, occurs organismically. The only alternative to this growth is decay, known euphemistically as development or progress within 'civilization' lifeway perspective and world view. This law means that human social, spiritual, and sociocultural growth like biologic (physical) growth occurs in timed sequential stages. This cannot be refuted by science since sociobiology has established the intergrowth and interfunction between biological and social growth, and since anthropology has established the intergrowth and interfunction of these with sociocultural growth. It means that realizing this—a crucial law of our condition of human authenticism—we must now understand the elements-phenomena of human socioculture as either growth or as decay; that is, as either the growth and perpetuation of human nature, being, spirit and reality or the stunting and/or decay of these. It means that we can only have natural socioculture that benefits a natural social nature and being that is timed, sequential stages of growth wherein each stage is maintained within the later stages of growth and within the final product, socioculture of natural, authentic human selflifeworld—nurtureculture—intergrown with natural Earthlifeworld (biosphere).

# 12. Twentieth Century History and L. Mumford

It is reasonable and useful, if one is to add to, enhance or reform the prevailing view of an age, to evaluate the very best thinkers of that age. To illustrate the breakthrough that authenticism makes from the reigning view, let us look at one of the giants among the American Twentieth Century thinkers, Lewis Mumford. It is hard to find a contributor that has surpassed Mumford in his time. He partially broke through the view of man as 'civilization' being to man as human being. Every great thinker, though, retains blemishes from membership in place and time. (Mumford, Lewis. *Interpretations and Forecasts: 1922-1972*: New York: Harcourt Brace Jovanovich, 1982; single quotation marks are added:)

> "Civilization itself had been a formidable invention, as costly, as dangerous, as the original Promethean gift of fire. Civilization was founded on the astronomical calendar, on written language, on the higher division of labor, and on the translation of habits and institutions into permanent buildings, monuments, cities; it marked a bold departure from the 'fossilization' of tribal societies: a gain in 'freedom', an intensification of life, which might be paid for in a disintegration against which the 'primitives' rigorously protected themselves. Static tribal societies were closer to the heartwood, farther from the cambium layer where growth takes place."

From human authenticism's view, two ideas here are striking. As an historian first and philosopher second, Mumford is willing to accept this trade-off: "a gain in freedom, an intensification of life, which might be paid for in disintegration," which we and Mumford know very well civilized society is repeatedly subjected to—its families, communities, states and sociocultures. He wrote these words in 1944 before the August 1945 atomic bomb destruction of Hiroshima: that is, before the "might be" took a quantum leap to the instant destruction of whole cities, and theoretically the rapid destruction through nuclear world war of the historian's 'civilization' itself. Further, with the conception that other complete destructions, from both warfare and environmental abuse, the final abyss appears, the end of all human life. Thus, this trade-off for humankind offered by 'civilization' suffered a quantum drop in desirability, with the loss being not just various, periodic disintegrations of particular sociocultures but the loss of all 'civilization' and any greater freedom it professes to harbor. Moreover, the apprehension of biospheric destruction brings the loss to one of all human life. This same Mumford in 1968 finally came to echo—what the youth of the sixties youth movement were also saying—the words of John Ruskin "'There is no wealth but life'— Let it flower."

Thus, history, 'civilization' and the historians are presently suffering a massive loss (or even total) of credibility. Attempts are being made to restore some of it by efforts to reduce the amount of destruction within nuclear world war to lower levels of destruction. But, with the mountain of sci-technic society's deity divided into two opposing camps, competition is the catalyst of destruction. Power justifies every pursuit which feeds it, in the name of self defense and survival; recently through a balance of power and unacceptable levels of destruction in war, as deterrent. In such a lifeworld all human values are now subordinate to deterrence/survival power. A reduction in destructive power is a trade-off: a return to acceptable mutual destructive power increases the chances that this acceptable, mutual destruction will occur. We do not need, however, nuclear weapons, nor even opposing camps, to envision the very real possibilities of societal and human disintegration, since disintegration occurs from within and without. Several other manifestations of modern science-technology such as biospheric destruction will suffice. Credibility becomes a scientific Humpty Dumpty of modern 'civilization'.

# 13. Futurism

Americans are characteristically optimistic about the future. Most of us spring from ancestors that were offered seemingly boundless promises in a newly discovered continent. Both the land and its native peoples were material to feed our asserted 'manifest destiny'. So much opportunity for so long fosters and breeds optimism. Western science and technology, united with the American, political system, represented a Western thrust of domination and subjugation into the loins of Earth-folk humanity. Many Americans were among the victims of this dynamic of power and wealth.

It must be remembered at the outset that the future actually does not exist. That is, it will never exist as existing phenomenal reality: but can only 'exist' as projected or prophesized conception; 'reasonable' semblances of some predictions about the future will come to exist only within ones limits of 'reasonable'. It is even arguable that no even modestly thorough visions or aspirations of future society ever have come to pass within any reasonably accurate approximation. We all can think of some prominent thinker that "would turn over in his grave if he could see what his vision of his system or idea has come to." If no plan, vision, or system of the future has ever been satisfactorily attained or a reasonable facsimile thereof, then the burden is on futurism itself as a quasi science to show that it is not an inherently false pursuit, as distinguished from unsystematized, single predictions. Further, it just may be that futurism is even a genuine, rather comprehensive escapism.

This Part and this volume show that not even a socioculture built upon the visions and hallucinations of hallucinogenic drugs could be any more escapist.

Human planning for the future has until recently been natural and valid; but, sci-technic society has seen the traditional form of planning take a quantum leap and become transmuted into futurism. The difference is that natural, authentic, human planning anticipates certainties or possibilities determined to be such through knowledge of events that have occurred in the past. The accelerating pace of innovation and change within sci-technic socioculture adds to anticipatable certain and possible events two elements: (a) the anticipations which are projections of new, present, sociocultural phenomena expected to either certainly or probably continue; and (b) anticipations of new phenomena still in the minds and/or on the drawing boards of innovators. Thus, planning for natural events that could or will happen is thereby transmuted into strategy or game planning aimed at bringing about phenomena that is in ideological dispute from two camps—differing innovators with differing vested interests and those disposed to established ways, i.e., to conservatism. Hence, planning becomes the ideology, futurism; the nature of the ideology is set by the particular vested innovation/interest. Futurism, then, is an ideology (rather than the science it would have the masses believe it is) which reigns over dehumanized sci-technic socioculture through its built in dynamic or activation mechanism, unquestioned, democratically unevaluated, innovation. Moreover, futurism also utilizes the doctrine of open-ended human potential or possibilities; and, thereby attains to a philosophy, alienism, intertwined with the power-wealth spirit-ethos of sci-technic socioculture. This philosophy furthering the frontiers of sci-technic 'civilization' is most distinctly marked by artificialism. But, the term alienism is more fitting as it stresses the realities we must stop trying to escape from: (a) that the nature of our human being is totally natural; (b) that we are natural, human being, nature, spirit and reality with a natural selflifeworld authentic to us; and (c) that all artificials are alien to us through the revelation that we have a purely natural 'culture' (socioculture)—nurtureculture—that is authentic to, fitting and belonging to us; and, that all artificials that have penetrated and adulterated this natural nurtureculture are adulteration and decay within our natural self and lifeworld, authentic to our 50,000 year old human subspecies. Failure to realize the quantum leap and transmutation, of future planning into ideology and philosophy results in its acceptance as modern, scientific, advanced planning, which it is not. Instead, it is still another escapism, our megaescapism, produced by socioculture trapped in unreal, incomprehensible, unmanageable conditions and forces; it is escapism attained to ideology; it is ideology recently attained to philosophy, the philosophy of alienism.

In original, authentic, human self and lifeworld and socioculture the future comes into existence as the present through the self-perpetuation of natural, human socioculture: the past was perpetuated and reproduced into the present and the present in like manner pulled the future into its archetypal patterns of living, human reality, being, and nature. The future was under general control, i.e., held within the evolution of self-perpetuating, human, subspecies, socioculture nature and biological living nature. The variations differing any 'present' of our subspecies temporality from future 'presents' were incremental, like the slow change to colder seasons, different game, dietary, and clothing changes. The future was experienced within the present as, also, was the past. Some binding social elements of this organismicity (unity) are organic family and community, tradition, custom, and ceremony; the center of this unity is natural, authentic socioculture and self-embracement of human self and lifeworld: this is to say, both past and future are valued as good, fitting, beautiful and right toward the purpose of self and lifeworld—the growth and perpetuation of natural, authentic, human nature, being, reality and spirit. This continuity is broken by the alien penetration and adulteration of the natural evolved being belonging to/with and authentic to natural, human self and lifeworld and socioculture (nurtureculture) of the species or subspecies, broken by alien, artificial change in evolved species, nature, and being.

# 14. Cultural Movements

At the beginning of broad philosophical movements and the isms within them, these contain both the coalesced beneficial elements needed by the age or time of particular, evolving sociocultures; and, they contain, as well, overshadowed, deleterious elements which are being overridden by the primary needed ones, but which are growing as parts of the broad movement or ism. The latter deleterients increase with time, becoming a decay to the structure-function-dwelling—selflifeworld—of the sociocultural system. It is when these unrecognized defects of an accepted movement flood into a sufficient collective awareness that they are spasmodically renounced in a revolution of conceptual consciousness, sociocultural purpose and social action: a new movement coalesces. All movements, both broad and segmental, are flawed with deleterious elements unrecognized by adherents until the critical mass of revelation-spasm. They are like this because every socioculture expanded beyond family and communal village is itself in its living nature adulterated in most spheres by alien elements. Human intelligence is natural and efficacious only within its natural range, sphere, its structure-function, in its natural, human being and that being's natural structure-function-dwelling— its natural selflifeworld and socioculture. Removed from their natural socioculture into alien sociocultures—adulterated through penetration and influx of alien elements-phenomena—humans and their adulterated, diminished being and intelligence function within and create systems that minimize the subliminal megamisfortune and handicap of being swept out of the freedom of their true and authentic structure-function-dwelling and its socioculture into alien socioculture and its transmuted, adulterated being and 'intelligence'. 'Civilization' men are condemned to the prison of 'culture' and to conceive and act within the flux of systems and movements that are adulterous and alien to the natural being, spirit, and nature, authentic to our human subspecies.

# 15. Joseph Campbell and Mythology

In his book, *Transformations of Myth Through Time*, Joseph Campbell states that the West has failed to transform its myths through time into forms that meet the evolving social environments or sociocultural conditions of the Western sociocultures. However, he fails to come up with the basic reason for this failure in the West to transform its old mythologies into that which relates and is of benefit to the changing sociocultures. There are at least four basic obstacles to this transformation of myth through time; one of which is the central or nucleus obstacle. First, as the number and population of states or nations increase, so do wars which often destroy the vitals of a mythology along with its socioculture's absorption or conquest. Second, the accelerating pace of change, especially within the Industrial Revolution, eliminates the time needed for mythology formation. Third, the intermixing of several sociocultures is invariably destructive to their mythologies, sometimes thoroughly so.

The core reason for this failure, however, comes from the nature of mythology. In ancient socioculture and in pre'civilization' socioculture even more so, mythology deals with, interprets and gives meaning to the natural phenomena and natural elements of living. These traditional mythologies related the natural elements-phenomena of the lifeworld—selflifeworld—to something which is also natural—to natural, human nature, being, spirit and reality. Mythology is essentially and largely a natural phenomenon which serves natural beings (human), their nature, being, spirit and reality by relating these to natural elements-phenomena around them that inspire awe, mystery and wonder. It is the very nature of mythology that it be an interaction between these two natural phenomena, i.e., natural humans (beings) and *natural* elements-phenomena. It is for this reason that as Western sociocultures have moved deeper and deeper into artificial socioculture, into artificialization, i.e., moved deeper into 'civilization', it is inevitable that these artificialized, adulterated, unnaturalized sociocultures are unable to create true and authentic mythologies to meet the evolving social environments and conditions of these sociocultures. They have become too artificialized, too unnatural to be able to create anything but phony, attempted, surrogate mythologies which do not take root because they are not natural and authentic to the natural human beings— the sociocultural environment is no longer natural enough to produce a viable, meaningful, relative mythology (metaphorical representation and expression).

Campbell's call, then, for a mythology that fits the sci-technic sociocultures of the West is futile and impossible because he speaks of a mythology that can relate a largely and progressively artificial sociocultural environment or lifeworld— a *quasi* life, world and self—with a natural, authentic human being, nature and spirit: this would be a nonviable, destructive, adulterative relationship. He speaks of the impossible; for, one cannot make a would-be 'mythology' out of artificial, sociocultural elements-phenomena and then apply it to, relate it to a natural, human being nature and spirit, without this being a mutilation/surgery upon human nature, being and spirit. This amounts to the creation and infusing of artificialism into human naturalism (human authenticism), i.e., it is the adulteration and decay of our natural, human authenticism, authentic, right, fit and belonging to us.

Human being, nature, and spirit through mythology responds to, relates with—intergrows/interfunctions with the natural, authentic, human selflifeworld. Mythology is one phenomenon which actualizes or stimulates human nature, being, spirit, reality and their selflifeworld. Mythology participates in the growth and perpetuation of our own natural, authentic structure-function-dwelling—selflifeworld—of human nature, being, spirit and reality. It must be natural; otherwise, it cannot participate in this growth/

perpetuation—it can only be adulteration, diminishment and decay of this growth/perpetuation.

Joseph Campbell also speaks of the warrior within us, that portion of us which struggles against the forces that would interfere or keep us from following our own true path. He is, of course, talking about individualism and affirming it in a new interpretation and portrayal of it through his perspective of mythology. In excavating and reviving the power of myth, Campbell's philosophy is a mythological neo-romanticism. He is addressing afresh the individual's struggle against the manifold collective forces within the city, state, nation, empire and 'civilization'. This story is of an origin of general compromise itself. Compromise with new demands presented by emergent, artificial collectives which are oversized and, which within their process of emerging formation, break off and disintegrate portions of the collective whole self, and then adulterate it through artificial, surrogated reconstitution into bio-intellect-egoself, and through integration with quasilife and quasiworld adulterated by alien, artificial elements-phenomena. But, there is and never has been a need for compromise within natural, authentic, human socioculture which, before and after 'civilization', has been outside of compromise and its alien, artificial, sociocultural collectives. In natural, authentic, Earth-folk socioculture there is and has been a natural interfunction/balance between the portion of the self which is ego and those portions which are family-self and community-self. Further, this natural, human organismicity that is compromised includes natural lifeworld, which includes both the natural nurtureculture authentic to our subspecies and natural environment authentic to all natural species.

Individualism is a song, a protesting, proclamational conception and doctrine, sung by the egoself against the emerged, artificial, alien collectives—against their damaging intrusive forces—which, it knows subliminally, have pulled it out of its natural constitutional formation—that of whole self intergrown with lifeworld. This song brings into awareness something that has been lost,

partially recovers it, and resists the presence and force of the penetrating, fragmenting, destructive, alien, artificial collectives. This
idea and doctrine song of individualism, along with the wider
philosophical song of romanticism, serves like song itself—as a
resistance and clutched affirmation of human being, nature, reality and spirit.

## 16. Living Nature, Intelligence and 'Civilization'

It is not just living nature that is smarter than sci-technic man.
It is natural, a'civilization', Earth-folk socioculture that has more
intelligence than the exploding, chaotic, perception and intellect
of sci-technic 'civilization'. Is it really a shortcoming, a sin, a failure for humans not to become gods! The truth is the opposite: the
most intelligent and successful thing any natural living being can
become is itself, to fulfill itself—to realize a perpetual fulfillment
(with each generation) of its living nature, of the living nature of
its being, spirit, identity and selflifeworld.

The greatest failure, sin and tragedy, is the intoxicated sci-
technic perception that humans and humankind have no boundaries to their activities; that they are infinite, open-ended beings
via some mystical, foggy, sci-technic, pseudodivinity; that they
have no sociobiological and sociocultural natures with boundaries,
archetypes and models that fix human being to just one star, one
shining magnificent light of warmth, growth and fulfillment—
the light of human spirit, being, living nature, reality and the
human selflifeworld wherein these are structured/functioning/
dwelling.

One of our boundaries, models and laws also fixes this light to
an enduring continuity open only to the very slow change of
Earthlifeworld (evolutionary) conditions; our endurance and perpetuation through time is not open to the very swift change of sci-
technic 'civilization's' artificial objects and phenomena; for, these
artificials, to natural human being and its life, mind and spirit, are
adulteration-pollution that we now realize makes natural human

beings debilitated through pathogenesis as it does living nature (the environment or biosphere).

# 17. Living Nature, Intelligence and Being,

Matter *per se* doesn't matter. What matters—has real meaning for us—is naturally perceived matter, which is participating, by the evolved nature of our human nature and being, in our natural, authentic, human selflifeworld, and which is known through natural perception and intelligence to be participating in Earthlifeworld (biosphere), and all non-matter that is naturally perceived by the natural consciousness, authentic, right, fitting, good and belonging to/with us as conscious, natural beings. Artificialized perception is diminished through its dilution/pollution with the adulterous artificial. And, since totally artificial being doesn't exist, artificialized being is diminished *and* alienized, quasi being.

Substantially artificialized being is substantially diminished being. Whatever aliens from another planet's living organismicity that may perhaps theoretically (a debatable theory) reach us would be our inferiors as natural beings because they would have to be highly artificialized and diminished of their natural being, the only being that exists. Its far more likely—even certain—that such aliens would self-destruct through adulteration and artificialization of being during the early stage of space exploration. Such a self-destruction through artificialized science, our own human species has started headlong toward.

How is it that so many scientists have placed inanimate matter at the center of reality instead of the living, human being consciousness: our consciousness that defines and experiences our reality? How is it that the great majority of scientists are defected, converted and absorbed by the artificial realm? Is it not an inherent fate of artificial science, if we do not give it a natural guidance system of human wisdom, that its scientists and their followers (engineers and the sci-technic population) are doomed to a progressive absorption into artificiality and its self-destruction!

Reassurance is ours; for, it is inherent to the natural, authentic, human condition—human authenticism—that it will experience the human revelation and a regeneration of natural, human intelligence, spirit and being to reverse this movement into artificial alienism. The only question is, how many will be able to.

## 18. Freedom Vs. Determinism

All living being is free including human being: natural being, in its various forms is held by living nature through natural selection and mutation—through evolution—to its living nature and being. Earth beings are free to be themselves. Yes, all animals are born free until humanity partially violates their nature and natural function through domestication to serve human needs and wants; likewise, we humans are born free until we violate our nature, being, spirit, identity, reality, and the natural nurtureculture, the nurturome, of these. The mechanism or dynamic that secures human beings within their living nature, that perpetually confines and keeps them in line with their freedom, experienced an original penetration and breakdown about 5,000 years ago (See HUMAN AUTHENTICISM VS. ALIENISM, Ch. VIII.). At that point the will of humanity's mind first turned against and partly opposed the freedom of being.

Prior to this emerging opposition of human will to natural, authentic, human nature and being, human freedom was determined—secured by natural mechanism; freedom and determinism were unified as one, as natural, evolved, unfolding, human being.

It is only after we fall that the will to get back up or rise occurs. Adulterated and penetrated humanity's off-balance, unnatural stride provides for the ongoing fallings from itself. City people stagger through life unknowingly; for, they were born into their alien, drunken state and call it living. It is within the nature of people and creatures to experience fallings and risings until a fall incapacitates us. After all, natural being is determined by evolved genes and nurtureculture to function freely: its freedom is determined through the living nature of natural being. However, it is not within living nature and determined freedom for a species, including ours, to fall outside it's nature and being. Every species maintains and perpetuates it's nature and therefore it's place in living nature, or is, otherwise, declining toward extinction.

It is human nature to nurse injured family and community members until recovery or death. There is, however, no Earth-folk medicine on the species level. In order to save modern humankind from falling outside its species nature, we must reverse the movement away from our species' nature and being, back toward these and our authentic, fitting and proven nurtureculture. Our choice is to either substantially and gradually recover our natural, authentic, human nature, being, spirit and reality, or to become decimated and/or extinct. Human authenticism is the species medicine, theory and practice unified, for the recovery and preservation of humankind.

Free will versus determinism, then, are false concepts to original, natural, authentic humankind. They both arose, as with Hegel's dialectic and the dynamic of alienism, with the departure of humankind from living nature and human nature. Authentic, natural human nature secures our freedom as determined by ecologically evolved, living nature and human nature *to* us. Authentic human freedom is natural, authentic human, nature, being, spirit, identity and reality, with their selflifeworld and nurtureculture within Earthlifeworld (ecosphere). Without the alienization and adulteration of 'civilization', our original, determined freedom would be secured by our dual nature of innate, genetic nature and nurtureculture nature. 'Civilization' partly denaturalizes us, leaving our nature, being and destiny, in part freely determined by our living nature, and in part cohersively imposed and determined by unnatural, unauthentic human socioculture. After 45,000 years of natural and free living, our human subspecies tumbled into a socioculture part natural and part unnatural, part real and part alien and artificial. The individual finds himself holding the resulting synthesis, and debating with himself about whether that which he holds is cotton or polyester, not knowing it is a blend of cotton and polyester. 'Civilization' life becomes like the blended cloth and other materials in our lives—partly of our natural, authentic human will and being, and partly determined by accumulating impositions of unauthentic, unnatural, alien socioculture. The will of 'civilization' man splits into a growing opposition, as will divided against itself: natural, authentic, human being and selflifeworld versus fragmented alien, artificial, quasi being and quasi, egoself, 'life' and 'world'.

There can be no 'free will' in natural, authentic humankind: for, human nature and being are determined by inherent genes and inherent, perpetuating, natural species' nutureculture. Humankind's only freedom of will can be that to oppose its living nature—to be self-destructive. True freedom in living nature is the growth and fulfillment of the organism's living nature; in human nature it is the growth, fulfillment and perpetuation of one's human nature and being, one's personal family and community nature, and being that one is evolved ecologically through genes and inherent nurtureculture, the being that is freely bestowed and secured by inherent genes and bequeathed and secured by natural nurtureculture authentic to our species. Only 'civilized', unnatural, adulterated humans are presented with an accumulation of elements-phenomena of unnatural nurtureculture (socioculture); and this is their determinism: that some of these anomalies cannot be escaped, which they "freely" surrender to. Their imprisonment in anomalous and alien elements-phenomena determines that they will acquiesce to that order. The variety of their anomalous elements are often accompanied by a freedom of choice, creating the illusion of free will that is, in reality, a freedom to choose from various alien, anomalous elements of the imprisonment outside their natural selflifeworld and its nature and being. Freedom of burgeoning choices still leaves imprisonment outside natural, authentic, human selflifeworld and its human nature, being, identity, reality and spirit.

There is no "free will" in natural, authentic humanity and its selflifeworld; there is only natural, authentic human will, will to be, will toward human being, its nature, identity, spirit, reality, and selflifeworld. Nor are there any 'determinisms' for natural, authentic humanity. Free will and determinism are a dual phenomenon; both arising only with emergent, artificial, alien elements-phenomena and flourishing in alien, unnatural, unauthentic socioculture of 'civilized', quasi egoself, 'life' and 'world'. Natural, authentic human freedom is secure, bequeathed, inherent being, nature, identity, and reality of authentic human selflifeworld. Only when elements of selflifeworld diverge and fall outside of natural, authentic nurtureculture into unnatural alien nurtureculture does man ask, "are these elements mine by my free will (choice) or determined for me?" The answer is they are both. Their presence is determined by historical force; the rise of 'civilization' accumulates unnatural alien elements-phenomena of selflifeworld and nurtureculture. Their presence brings the emergence of unnatural, unauthentic will, of 'civilized' (alien) will, which assumes two functions. First, the variety of form and their degrees of stress and pain pull the human will to being and to nature identity and reality into a new unnatural alien function: to struggle for at least the freedom of choice—it wills toward the least stressful of these presented alien elements; it wills toward autonomy, toward self determination (selection) from determined choices, and away from random or outside imposition of these elements. Second, the mere emerged presence of alien elements amidst natural, authentic elements of human nurtureculture and selflifeworld divides the human will with the natural, authentic portion willing toward being's authentic freedom; and the new alien portion willing toward freedom of choice.

## 19. Freedom and Automatized Sci-Technics and Commercialism

There is something in the human psyche of modern 'civili-zation' humans that, when science-technology produces the possibility for doing something, this is immediately followed by the conclusion that this possibility almost certainly *should* be done. In other words, that once the means have been cre-ated for doing something adventurous, novel and exciting, that this very creation of the means, through our ethics of adven-ture-discovery-novelty, certifies that the undertaking is fitting, proper, correct and good for humankind to pursue. In its way with humanity, modern science-technology routes normal, natural, authentic human value judgment and replaces it with an automatic acceptance or seal of approval given to possibili-ties that are exciting, adventurous and challenging. How and why has this happened; and what does it mean?

Two basic things, among others, have happened. The adven-ture-discovery spirit-ethic of the West's age of discovery that was interlinked with a subjugation-exploitation ethic was, more re-cently, interlinked with a science-curiosity ethic and a technol-ogy-challenge ethic, both having permeated the masses through radio and TV, forming a sci-technic adventure-discovery spirit-ethic. But there is more involved here than the retransmutation of an earlier, Western spirit-ethic. There has also been a change in the sociocultural manner or process (aside from the socio-psychologi-cal) through which science and technology enters the phenom-enon of socioculture. For, this retransmutation is one that enduces a human self-deception resulting in the abdication of our genuine, evolved, human purposes to those of a self-directing—automa-tized—mechanismic socioculture. Thus, a new kind of 'manifest destiny' has captured not merely the American Indian within the clinched fists of the machine and its science, but also the West itself within the grip of automatized, mechanistic, sociocultural dynamic and development. Man is conquering living nature with the machine only to eventually find the machine is conquering him. All is fair in power play.

The vital human options essential to human freedom, i.e., those of available kinds of work and available kinds of leisure activities, and consequently, the very ways that we relate with ourselves and others, and with objects-phenomena within 'civilization' socioculture; when these options are created by various, integrated, mechanismic processes, when these processes become automatized—self-directing—and when they automatically create new automatized, mechanical operations driven by automatized processes and produce work and leisure activities automatically determined according to the nature of the unnatural requirements of this general automatization, then the freedom to choose from among all of these activities that have been automatically produced by the values of automatized mechanism, rather than produced through human values, this freedom is merely scraps of freedom: whatever of human spirit, nature and being that remain yet unpulled into an automatic, human being grinder. In this way automatized personal, social and sociocultural change, movements and developments, as well as behavior, activities and projects suspend human value judgment—indeed, the moral faculty itself—replacing these with an automatized seal of approval. The tragedy that R. W. Emerson recognized, that "Things are in the saddle and ride mankind.", has become the tragedy that automatized machines and their science-technology with their self-directing, integrated systems are riding mankind into a progressively inhabitable, self-destructing, alien, artificial jungle of objects-phenomena and human behavior.

This is the tyranny of automatized mechanisms, a tyranny equal or greater than the traditional tyranny of a dictator or a small group or class of people; for, the latter type of tyranny can be quickly overthrown by the tyrannized masses. For, when a person or group of people are on our backs, some or most of us, despite religious or political propaganda, are naturally aware of it through our remaining natural perception and intelligence. However, when things are in the saddle riding people, sociocultures and a large part of humankind, this fact defies our true, natural perception because of the fact that we are operating the machines: from this we get the impression that we are in control of the machines. Never the less, once 'civilization' people come to realize that the artificial 'will' or 'purposes' of the machine are directing and selecting the work and leisure time activities and degenerating interpersonal relations, family and sociocultural functioning of that humanity within 'civilization' lifeway, rather than guided by the natural, human will, purposes and values that are authentic, right, fitting and belonging to us, then 'civilization' peoples will be ready to plan and execute a revolutionary movement as important as the historical, revolutionary movements against human tyranny.

What a bitter trick history has played on us in our struggles against tyranny. Just when possibly the finest form of government for nations was being established, i.e., the democratic republic of the United States of America, a new kind of tyranny was taking hold within the sociocultural body of Western 'civilization' socioculture. It is almost as though—and probably is the case— that the added burden of this new oppression of human freedom, though still unrecognized and masquerading as a great new benefit to humankind, was felt and recognized at a deeper level within the peoples' spirit and being, forcing them to make a breakthrough to largely overthrow the older burden, the long established form of tyranny, human oppression. It is true, then, that it is precisely at the time and place of the flourishing of modern socio-politico-economic freedom that human freedom itself is slipping away amidst these flourishings—even while we alertly protect and participate in that complex socio-politico-economic freedom, that human freedom itself suffers a new, equal deterioration from the automatized forces of science-technology; forces which are automatically creating new, alien activities, behavior and lifeway elements-phenomena that are generated automatically by integrated, automatized, mechanismic systems that are almost totally free from human value judgments through our unawareness of the antihuman, imprisoning nature of these automatized, mechanismic systems.

## 20. Computers

The most extreme and fullest form of the artificialization that humans have created at present is the computer. Computers are increasingly capable of performing calculation and 'intellectual' kinds of mental abilities that, in Industrial 'civilization' socioculture, some humans recently find necessary or expedient to perform. But, it is a mistake to presume that, because a very small minority of our human subspecies have very recently began performing these kinds of 'intellectual' activities, that they are by that measure human activities. That is, just because a minority of humans do something for years, for an age or for centuries does not necessarily make it characteristically human. To the contrary, abilities that have only been very recently acquired by a very small minority of our human subspecies cannot possibly be defined as characteristic of or placed within the nature of our human subspecies—cannot be placed within the realm of human nature and human being.

Humans are the only (natural) species capable of imagining or dreaming up objects or activities that are outside of their nature and being to engage: and, then, to engage them because it is expedient at the moment or within the time or age, and because the condition(s) of their unnatural, blended, 'civilization' sociocultures make it expedient to do so. These sociocultures are partly within human nature and being and partly artificial–partly outside human nature and being. The human imagination can be used naturally, i.e., to imagine and engage objects and activities that serve natural, human purposes authentic to our species and subspecies; or, it can be used unnaturally by utilizing artificial objects-phenomena that have penetrated the mind for the imagining and engagement of more artificial objects-phenomena. The imagination, however, is natural and must be penetrated by alien, artificial objects-phenomena in order for it to utilize them for expedient, unnatural goals stemming from desire for power and wealth outside natural purposes authentic to our natural subspecies.

Ultimately, this ultimate artificial creation will demonstrate to the few that are capable of realizing it (and an extreme few are presently breaking into this realization) that artificial objects-phenomena themselves no matter how sophisticated can never be fit within, and be beneficial and wholesome to the human realm which is an organismic integration of natural, living objects-phenomena and natural, living organisms—intergrowing, interfunctioning and interliving; can never be fit into the organismic ecosystems of life, which the science of ecology now calls *ecosphere* or *web of life,* and which, here in this volume and this philosophy, is called Earthlifeworld. The lesson will be that the purpose of human endeavor and living is not to create progressively sophisticated forms of artificial abilities: rather, it is to be conscious of the natural, authentic purposes of our natural, authentic, human nature, being, spirit and reality; and how they fit and play organismically within human selflifeworld, which, in turn fits, interplays or interfunctions with the megaorganismicity of Earthlifeworld; which is an evolution-growth-perpetuation. The computer is, then, ultimate, artificial, adulterant-pollutant-destructant to natural life's growth-fulfillment-perpetuation of humanity.

## 21. Cybernetics

Cybernetics–by accentuating the limited comparables and pushing aside the contrastables or incomparables between the artificial, dead machine and the natural, living mind and being–is the institutionalized salute and homage to the machine and the insult, assault upon, and dishonor of our human mind, being, nature and spirit. This is a major part of our human self-destruction; this glorification of the *activated* inanimate and this ready lack of spiritual-ontological allegience to the glories of human being, mind and consciousness. These obtuse pseudo-human scientists should realize that it is their own heads that require their studious examinations—their minds, spirits, values and identifications need to be examined.

The lumping together of the control and communications of the machine with that of the mind is a step backward into a new shadowy darkness for humankind. For, the first and most important fact to be apprehended about machines and mind is the definitional distinctions dividing their inherent natures. However, physical scientists and mathematicians enamored and devoted to their diciplines start from the opposite direction and place comparables of—instead of definitional distinctions between–(a) ecological, organismic nature as the foundation for any genuine, sustainable system of intelligence, and (b) an automated, artificial system; thereby, they are assuring at the outset that the dicipline is basic, human folly, despite whatever bits and pieces of inappropriate, dehumanizing knowledge feed their delighted

curiosities, i.e., feed their addiction to and deep belief in curiosity, discovery and invention.

## 22. Heidegger

Time after time Western man's pursuit of the elusive real self and real world has been cut off by the 'missing link' in Western understanding or perception—the key insight that would rejoin him to both his real and natural understanding and his real and natural lifeworld. Heidegger, in his philosophical journey, developed his concept of authenticity just about as far as one can lacking this missing conceptual link. In fact, he knew that Western understanding and consciousness was a false one and felt that we could prepare ourselves to receive a revelation that would bring a new understanding, if and when this might happen.

It is important to underline at the outset that human authenticism demystifies and unesotericalizes the pursuit of authentic, human self, authentic human reality and being. It does so by doing what Heidegger could not—by breaking through to the revelation of a natural, authentic human nature and being composed of a natural, authentic self intergrown and interfunctional with a natural, authentic lifeworld (social units and environment). In the later part of his pursuit, he came as close as he could, being trapped behind the barrier of the existentialist's blindness to the facts of our natural human nature and being authentic to us, and overly confined to the spiritual-ontological grounds he and others landscaped diligently.

Heidegger holds that human existence constitutes the openness where being can be revealed; for something "to be" means for it to be revealed, uncovered, made manifest. Authenticism, instead, holds that authentic human life constitutes the structure-function-dwelling where natural, authentic meaning(s) and reality indwell and are experienced. Heidegger's idea that 'authenticity' means to be most appropriately what one already is—as the open or empty self freed from egoism—is replaced by authenticism's assertion that to be the authentic self is to embrace and dwell within a larger realm than that of ego open to quasi life and world; that is, a realm of tri-ecologic self interfunctioned and intergrown with lifeworld. A realm that is naturally open to all the human being natural and authentic to it, and intergrown with all other natural beings—with Being. This living, human realm is naturally open to and full of all natural elements-phenomena that are authentic to and belong to/within it, and closed to alien, artificial elements-phenomena—it has a perimeter, a boundary. This structure-function-dwelling consists of natural, authentic people, pursuits, endeavors, thoughts, feelings and values; of natural, authentic family and other social units, behavior and socioculture. The philosophical system of human authenticism traces the source and makeup of 'unauthenticity' (and 'authenticity') to deeper and wider realms than Heidegger's concept of 'inauthenticity' as egoism and its manifestations. That is, just to be free from the egoism of egoself is a starting base only and is insufficient and simplistic.

Heidegger arrived towards the end of his journey at an understanding and a kind of awareness or enlightenment that others before him arrived at earlier in the journey of understanding and life. In America, Emerson arrived at the concept of the 'over-soul' at about age forty-five. And, as Lewis Mumford (Mumford, Lewis. *Interpretations and Forecasts 1922-1972*, New York: Harcourt Brace Jovanovich, Inc., 1973.) points out "Thoreau picks up where Emerson leaves off". We can assume that with the publication of *Walden* in 1854 that Thoreau, at age thirty-seven, had refined the American, transcendentalist version of, and his own experience of, the enlightened awareness we are speaking of. He, however, like Rousseau and the 'later' Heidegger was confined by historicism and one of its recent Western doctrines, individualism. They could not break through 'civilization's' paradigms and ideals, nor push beyond romanticism itself; Rousseau and Thoreau being held by pastoralism, Heidegger by industrialism.

In his early writings Heidegger believed that when our historical existence is authentic, our past is freed for the future by the resolve of the present. He did not realize that our historical existence is by definition, partly authentic and partly unauthentic. And that resolve is thereby incapable of undoing the organic and organismic decay and stunted growth effected by alien adulterations of selflifeworld and its being and reality. And, he never came to conceive of human authenticism, i.e., the complete realization and unity of *all* human authentics.

Presumably, then, a student or seeker of enlightenment, especially an American, has other more impressive teachers to follow, unless, it could be argued, he is careful to confine himself to the 'later' Heidegger (from Zimmerman, Michael F. *The Eclipse of the Self.* Athens, OH: Ohio University Press, 1981.):

"In his famous essay on releasement, the notion of releasement bears remarkable resemblence to the notion of enlightenment in Zen Buddhism. In this context *Ereignis* (Heidegger's sought after kind of understanding or awareness, Being) can be understood as analogous to the *Tao*. When Heidegger says that man is most himself when he acts spontaneously or naturally ('like a rose'), he shows his sympathy for a non-historical, non subjective understanding of the happening called *Ereignis* or Being, *Tao* or *Logos*. The authentic person, then, is one who is in tune with this happening: released from the artificiality of egoism, he is open for the natural 'way' of things."

Heidegger's first failure here is in not realizing that the "things" he speaks of—lifeworld things—are a blend of natural and of unnatural, alien elements-phenomena; that the egoself is only the first of the artificials; and that what remains is an unnatural "way" of a blended lifeworld—partly artificial and partly natural.

Heidegger in his search for authentic man, authentic life (being) and authentic, real, human life ("Being") came as close as he could but stopped short of seeing man as a part of nature. One of the powerful myths of our 'civilization' man overpowered even Heidegger—the idea that there is nature and there is man and, though having many relations and connections, they are separate realms. His Western, theological beginning would not release him to accept the chief unacceptable—that we evolved from previous forms of being, i.e., organisms, most recently ape-men, and retain in ourselves a part of them. That we are, when free, natural and authentic, totally of living nature. Again (from Zimmerman's, *The Eclipse of the Self,*): "The idea of Ereignis attempts in part to express the intimate correspondence between human existence and being." Thus, the authentic man Heidegger proposed could only intimately correspond with ultimate reality (in "Being"). He could never be a part of ultimate reality and live authentic human reality: human nature, being, identity, reality and spirit in a real, authentic, natural selflifeworld and nurturome of Earth.

The concept of 'authenticity' in the later Heidegger became married to the modern, Western concept of 'supermen' and restricted its offering to a very small minority of humankind. There was a resultant trade off: his philosophy gained in its understanding of enlightenment and the enlightened men in history, and lost any significant import to the masses. Thus, Heidegger in the end failed to discover and offer to the West what Zen Buddhism 3,000 years ago, Taoism, authentic versions of Christianity and other religions afterwards succeeded in offering humankind—an enlightenment of consciousness for the individual common man and his socioculture. Heidegger, if he had had a stronger realization and experience of human spirit (spirituality) would have been home free. This deficiency was linked to his deficient acknowledgment and experience of human nature; together these two human naturals participate in the phenomenon of human being; the experience and awareness of the three being interfunctional.

Part of Heidegger's limits is in his terminology. In fact, the concept and term, 'Being' that this ontology focuses on in order to understand and portray human life and consciousness is inadequate at the outset toward the portrayal of both human consciousness and human life—selflifeworld. We are more than beings; and our life more than being: we are human nature, being, identity, reality, spirit and their selflifeworld. Such it is that 'educated' men are so utterly lost by the twentieth century that the notions they pick to enlighten the way back home throw insufficient light upon the dark surroundings. Heidegger and all others, upon finding themselves lost within the strange, modern world, failed to ask 'just how lost are we?' and to sufficiently probe for and discover an answer. They failed to understand the extent and the exact nature of our lost condition. 'Civilization' men are long time suffers of incomplete diagnosis.

Heidegger's pursuit of the 'authentic' self and reality became lost, intertwined and confused with the pursuit of primal language. In the end, Heidegger ends up a latter day Romanticist. For, as Lewis Mumford has said, "the entire essence of Romanticism was an attempt to recapture the primordial". Heidegger came to the pursuit of some lost language in his pursuit of a sense of being, without knowing that both quests were the result of something else that he and 'civilization' man have lost—a portion of human being itself, and of all that is inseparable from it: natural, human nature, spirit, reality and selflifeworld authentic to these.

Heidegger and nearly all educated men start from a place so far from what human life is really like, so far from its essences, that they start on the journey of understanding it from within a darkness unknown to Earth-folk people (of Earth-folk socioculture). People generally within Earth-folk sociocultures—distant, recent and modern ones—unshattered by the penetrations of 'civilization's' alienism, intuitively grasp, experience and express human life through natural, authentic customs, ceremony, ritual, folklore, work and mythology, as well as through the meditations and revelatory waftings of authentic nurtureculture and the phenomenon of human life, but feel no need to dissect, study and define it with intellectual verbiage. All the while, thinkers/writers in the urbanvilles of sci-technic sociocultures realize the loss of and seek to define human life, its nature and meaning as a means to regrasp it, experience and express it. Such thinkers/writers are valuable to the multitudes in the West who are equally as lost as they. But, there is a danger to the reader who has, through his own independent insight or revelation, or by that inspired to him from enlightened writers has come to a stretch on the road where light has broke or is spreading. For such a reader a pitfall threatens—the assumption that every recognized thinker writes from a view more enlightened than the reader's own—and this can lead to confusion and even regression.

Heidegger's concept of 'authenticity' was to remain a concept within his linguistic, existentialist, phenomenological ontology system. There are at least four central reasons why Heidegger could not break through to the new system of human authenticism outlined herein. First, Heidegger, influenced by linguistic philosophy, made the inquiry into language one of the pivot points of his philosophy. Heidegger's pursuit of true self and true reality became intertwined with and confined by the pursuit of primal language. He was the spirit of Romanticism breaking in upon primal language. Second, Heidegger was captured and confined by the individualism still reigning in our age. He addressed what he saw (correctly) as a crucial problem for the individual—the takeover of the entire conscious self and human being by the ego and the resulting lack of enlightenment about all that is beyond the ego; but, he himself left out any significant account of the immediate social units, i.e., family, extended family, clan, tribe, village, and community. The world for Heidegger was the individual, other individuals, things, society at large, time and being. His central applications go to individual and mass society. He saw overwhelmed individual and overwhelming society. Being of his age (though in later life partly transcending it) he took both microscopicized perception and novelty, applied the Romantic spirit and commenced to innovate a focus on 'Being' as reflected by primal language. His third central limitation, then, was the inherent, paradoxical flaw of Romanticism itself: that it uses modern, unnatural tools in its effort to recapture primordial, natural essences.

Deeper still was the most central limitation upon Heidegger blocking him from the breakthrough he felt may eventually come to free Western man, i.e., his existentialist denial of human nature; this limited his understanding of authentic, human being which is natural and devoid of the artificial until adulterated by 'civilization' socioculture.

Further, Heidegger like Rousseau and everyone else in the West that perceives man as overwhelmed by society, advocates that he flee back to himself; for Heidegger, to individual being. Individualism is one of the two classic remedies for alienation; the other being statism or nationalism. (A major, recent solution, Marxist communism taking exception in proposing—but then failing to deliver—a social sea of brotherhood as the 'community' for the individual.) But, authenticism offers more than self-sufficiency; it offers social and spiritual sufficiency and fulfillment—the natural, authentic individual within natural, authentic social units and natural environment.

Finally, Heidegger was Romanticism merging with an spiritual-ontological metaphysics that was a realism merging with an idealism. Heidegger delivers up an existentialist, phenomenological neo-idealism.

# 23. Symbolism

The use of symbols is one of the characteristics often used to distinguish humans from pre-human animals. However, for a characteristic to be a distinguishing characteristic it must be present to a general extent and for the general duration of the species, genus, etc., in question. There is no reason at present to believe that symbol use occurred to a general extent in the early human species, Homo erectus, emerging about 500,000 years ago, nor in the species that followed it—Homo sapiens—and the first sub-species of Homo sapiens—Neandertal man. With the second sub-species of Homo sapiens which succeeded Neandertal man—Homo sapiens sapiens, to which contemporary man belongs, the use of nonverbal symbols did not achieve a general occurrence until the emergence of 'civilization'. Unless, that is, one is prepared to rely on burial of the dead and symbol use associated with hunting rituals and basic, human ceremonies alone to represent symbolism and uphold the thesis in question. Thus, symbol use is not even a general, defining characteristic of Homo sapiens sapiens, our own subspecies, and more certainly, not of our species, genus or human family; because symbolism did not achieve a general occurrence except in the socioculture called 'civilization' which only emerged 45,000 years into the 50,000 year old subspecies, Homo sapiens sapiens; commencing at that time to progressively radiate religious and politico-economic symbolisms outward from city lifewayinto rural life mode.

The written word and drawing are two principle types of symbols that are in use today by our subspecies. The first, writing, did not emerge until about 5,000 years ago, and even contemporarily the majority of human kind is illiterate—cannot read or write; therefore, it is obvious that literacy cannot be a general, natural, authentic, defining characteristic of our subspecies. As for drawing, if one assumes that drawing played a general role in human socioculture beginning in the early or middle period of our subspecies, then drawing type symbols can be considered a general and defining characteristic of our subspecies. On the other hand, if a general use of drawing in human socioculture cannot be found in a majority of these until 5,000 to 15,000 years ago, even drawing, then, fails to rate as a general characteristic of our human subspecies, Homo sapiens sapiens.

There seems to be at least two possible conclusions concerning our subspecies and symbol use having philosophical importance. The first is that symbol use such as in drawing, burial, ceremonial and ritual use reached a general occurrence in our subspecies at a sufficiently early time to rate it a characteristic of our 50,000 year old subspecies. However, this falls short of the idea that the use of symbols is a distinguishing characteristic of our 300,000 year old human species because no general use of symbols has been demonstrated as belonging to it or the human species, Homo erectus (about 300,000-500,000 B.P.). Unless, of course one wishes to oust Homo erectus from the human family, grouping it with ape animals instead of human animals. The second philosophical conclusion is that Homo sapiens sapiens is characteristically and distinctively a symbol using subspecies; but, that with the emergence of 'civilization' and the invention of the written word and other symbol use, symbolism pushed beyond the degree normal to the 50,000 year old subspecies and appeared along with the written word as an excessive symbolism outside of or beyond normal, natural human symbolism, i.e., an anomalous use of anomalous symbols. That is to say, for our 50,000 year old subspecies, writing and other (nonverbal language) symbols seen only in the recently arising form of socioculture, 'civilization' socioculture, are an abnormal, unnatural, alien use of symbolism, a quantum leap out of our human subspecies' natural, normal use of symbols authentic to us.

The significance of the first of the above two conclusions is that it agrees with the position of the philosophy presented herein—that the use of symbols cannot be a characteristic that distinguishes humans from other animals because of the mere fact that Homo erectus and any other preceding human species one may wish to acknowledge, did not achieve a general enough use of symbols (if any) to rate symbol use a characteristic of Homo erectus and any preceding human species. The significance of the second conclusion encompasses that of the first. Further, the second conclusion more thoroughly utilizes the revelation of human authenticism.

Ultimately, one has to go to our predominant, hunter-gatherer socioculture and the symbolic ceremony or ritual associated with it to support the thesis that the use of symbols is distinctively human. Even here the thesis meets increasingly heavy dispute as one moves back into the Neandertal man of our human species; and becomes very weak theory with Homo erectus and predecessor(s). Also, there is the question of when the use of a symbol or symbols attains to symbolism.

In summary, the use of nonverbal—nonlanguage—symbolism to separate and distinguish humans from animals is invalid. This failed attempt springs from such diverse stimuli and motivations as religion, egoism, egotism, arrogance, 'civilization's' arrogance, and the need for self-justification on the part of science-technology, 'civilization' sociocultures and religions, as well as, all the social organizations and institutions beyond organic family and community. (Language itself is addressed in Ch. X, Part 40, "Linguistic Philosophy" and Part 35, "Language;" and, sporadically, throughout this volume.)

## 24. Overpopulation

For humans and for other social species, overpopulation is not merely a matter involving security, food and water, and the activities of mating, bearing and rearing of young. Humans, along with many other social species such as primates and some mammals, have evolved another nature beyond the biological, that is, a social and sociocultural or nurturecultural nature. All human and many social species possess a 'second nature' consisting of natural nurtureculture—a natural socioculture authentic to their species;

and these natural, sociocultural, human species inherently are subject to natural, controlling mechanisms which control the population growth keeping it within the limits that their natural, sociocultural nature can bear while maintaining this nature. When population growth in a human species rises beyond its natural boundaries through the introduction of artificial means of food production which permit a biological multiplication of the population, this artificial rise in population—overpopulation—results in an incremental transmutation and adulteration of the natural socioculture authentic to the species. Consequently, the more ingenuity, artificiality or invention that is used to increase the biological numbers of the species, the more resulting concomitant violation of the nurturecultural nature of the species occurs. Therefore, an overpopulation beyond biological overpopulation—social overpopulation—occurs in the unnatural, quasi, human, sociocultural species, of 'civilization's alienism, Homo alienus; wherein the density of the human population in its natural living area pushes the socioculture outside its constitutional, social formations of nomadic or settled communities, i.e., outside the social boundaries of its natural, sociocultural constitution or nature.

## 25. Western Religions

Western religions are a great retreat, departure, separation from, and denial of living nature. Since human spirit (and all spirit) is necessarily natural spirit, just as human being is natural being, any religion that progressively moves humans into artifice, which artificializes human life, through its organizational, hierarchical and ritualistic trappings, is guilty of being a contradiction and counter productive to its avowed aim of increasing or cultivating human spirituality which is constitutionally natural. Christianity and Judaism assert in their commonly used Old Testament of the Hebrews, that man has fallen from living nature, from his original state of belonging with/to living nature and of being in harmony with living nature. Thereby, these religions lay the doctrinal basis for religion to save man from his fallen state out of living nature who was originally with living nature, and his fall from the "Creator" of living nature. Without this sanctioned separation from living nature, Western religions would have to send or fling man into living nature, instructing him to seek harmony with the spirit and essences of living nature. He would not, then, in that case be instructed to enter the artificialized institution of the church, the elaborate, ceremonial buildings of the church, in order to find communion or contact with the spirituality within him. Instead, humankind would be instructed, rightly so, to exodus and disperse out of the city away from the institutionalized, centralized, denatured religious centers and into the bosom and source of his spirituality, natural family-community village—extended family, clan or tribal communities, where he would be free from the complex, artificial structures of religion to receive first hand the simple, natural awe, wonder and inspirations radiant within the selflifeworld of a'civilization' (folk) socioculture that are natural and authentic to our subspecies of humanity. (In addition, the three great Western religions have quite consistently adopted the power-wealth spirit-ethos which, as is explained in Chapter II, is the central force beguiling and forcing people into 'civilization' sociocultures.)

In order to avoid the above nature of spirit contradiction, religions require another cardinal tenet also missing: that those people who cannot for practical reasons achieve a rural selflifeworld should be helped to identify, avoid and diminish the alien, artificial elements-phenomena that are blended into their 'civilization' sociocultures, and to increase the natural, authentic elements-phenomena: and ongoing movement of struggle is required, i.e., the struggle for natural human authenticism and against unnatural, unhuman alienism.

# 26. 'Primitivism'

Human authenticism is not a primitivism. To put it another way; what existed before 'civilization' was not human primitivism, but human authenticism. The concept denoted by the term, primitivism, is false and the phenomenon is illusion because the particular kind of human socioculture which the term, primitivism, has been used to denote has been falsely understood and used always, except by a small portion of anthropologists, in the critical context of 'civilization' socioculture centricity, perspective and bias. It was not the admirers of or the participants in these various 'primitivisms' that termed and conceived them as primitivism; but 'civilization's' critical judgment that did so. The basic, sociocultural phenomenon to which the term refers is, of course, prehistoric or pre'civilization' socioculture (though very early 'civilizations' are often included). This pre'civilization' socioculture is revealed by authenticism to be the natural, original socioculture authentic to the nature, being and reality of the human subspecies to which 'civilization' man belongs; but, from which he has fallen, into the adulterated, decayous, socioculture of 'civilization'. The term and concept, primitive, most often confusingly connotes beyond 'the simple and original' to include underdevelopment, immaturity, and that which has not yet arrived to the good, admirable or the complete.

This issue, then, hinges upon perspective and definition. The term and concept 'primitivism' arises from a 'civilization', socioculture perspective, centricity and bias. The shift from the historical to the anthropological perspective centered upon our subspecies, its past and present Earth-folk sociocultures, its natural being and reality, is a prerequisite awareness and revelation for 'civilization' man's final human revelation of human authenticism. Without it human enlightenment, both ancient and modern, will remain contained within 'civilization' socioculture's deluded self-consciousness which mistakes progressive, spreading decay for development, and contemporarily mistakes dead-end decay into progressive alienism to extinction/death for the illusion of doctrinary, open-ended development of human potential.

## 27. Anthropology and History

It has been traditionally acknowledged that we cannot achieve a highly objective, historical account of the human past approaching the objectivity we achieve in the natural sciences and mathematics. The thesis behind this assumption is that no one can escape from ethnocentricity, from the subjective influences and experiences of our time, age, city, state and/or nation, as well as, the centricity of a particular 'civilization' socioculture. However, the recent determination by anthropology of our human subspecies age to be at least 50,000 years, along with a substantial body of accumulating knowledge from anthropology, sociobiology, other human science, and ethology provides a growing body of generally objective knowledge of the true, natural and authentic human nature, being, identity, reality and spirit of ourselves as a subspecies. We know now that we as a subspecies of being (c.50,000 yrs.) and species of being (c. 300,000 yrs.) are much much older than we have previously and traditionally assumed and based our thinking upon; and, we know much much more about this longer past. These two realizations, along with the fact that 'civilization' only emerged at a few sites about 5,000 years ago and only as late as the mid or late twentieth century engulfed a majority of our subspecies into its urban lifeway and consciousness, provides for anthropology, together with the other human sciences mentioned above, to assume the role that history has heretofore played, namely, of establishing the archetypes, criteria and paradigms with which we are to understand ourselves personally and collectively, and to direct and manage the social and manifold, sociocultural units of our subspecies.

In effect, anthropology with the aid of the human-related sciences can now commence the absorption of the dicipline, history. But, this will not occur without a great struggle of resistance from history and all its subjectivities, methodologies, ethnocentricities, sub-branches, ideologies and manifold isms, as well as, from religions. This absorption of human history by the growing objective knowledge about true, natural, authentic, a'civilization' humankind, past and present, is furthered through the revelation of human authenticism, the revelation of our natural, authentic, human condition as it was and is now within us, albeit partially transmuted. And, inclusive in this revelation is the revelation of the condition of 'civilization' socioculture as a human condition that is no longer pure, natural and unadulterated, but instead is adulterated, diminished, fragmented, surrogated and disintegrated, and is to be seen as a decaying process and condition within our natural, authentic human condition of human authenticism.

The deliverance of humankind requires that we subordinate the adulterated subjective, and all the adulterated subjectivities (for, if we are within any 'civilization' socioculture, we are adulterated) to the unadulterated, natural, authentic objectivities, i.e., to the objective perceptions, conceptions and consciousness of our natural, authentic, human nature, being, spirit, reality and selflifeworld, and to the needs and purposes of these for their growth-fulfillment-perpetuation. Only the human revelation can assure that this ongoing dialectical struggle, now being intensified, between the human subjectivities and the human objectivities will now be one of a commencing and continuing triumph of human authenticism, as we change and reverse the traditional course of 'civilization' socioculture and move back into the condition of human authenticism through the process of human reauthenticization from year to year, decade to decade and generation to generation.

# 28. Religion and Science

One of the key failures and misunderstandings of the major religions is their failure to realize the great impact of nurturing upon infancy and early childhood. The religious teachers and leaders are educated men, and educated people have a need and vested interest in believing that formal education, which begins around age six, is the most important influence in attaining valued, adult attributes, whether religious or secular. Both secular and religious teaching largely share a basically false idea of the educated fool, that the learning that will determine character, personality, thought and conduct, largely begins at age five or six; when, in fact, the learning that forms these adult attributes has already largely taken place; and, furthermore, is greatly influenced by an even earlier phenomenon, i.e., the creation of genetic instruction influencing these attributes at the moment of conception. Thus, urban religion and secularism both have had to endure the major handicap of being largely ignorant of the roles of genetic instruction and pre-school nurturing. Resultingly, their different goals—spiritual revelation and secular 'success' at living, respectively—have all along been predominately pursued with formal education, as the general molding agent, which at best is one third of the matter.

Secularism, at least, has evolved a science, psychology, one faction of which argues for the correction of this traditional falsehood struggling against the faction of psychology that supports it. Religion, at least, has traditionally had some sects or denominations that value natural, non-materialistic lifestyles which result in the infant and young child significantly or largely escaping transmutational, alien, artificial and unspiritual elements-phenomena of urbanism, and especially of sci-technic urbanism. Still, religion is condemned to struggle within the major illusion that it can redeem souls that have been irredeemably lost or diminished of their spirituality. Further, at this writing it is clear that most other major sociocultures along with some minor ones are now following the West in its science/technologicalization which progressively generates ever new destructive phenomena that adulterates and diminishes infant, child and adult growth-fulfillment and perpetuation of human spirituality, consciousness and being, authentic and natural to us as birthrights. It is no wonder that so many Christian children by later childhood or adolescence can be easily convinced that they were born with 'sin', against the truth; that they were nurtured into it from birth onward.

However, the ultimate *spiritual*, human Protestantism is now in its beginnings. For, orthodox 'human' myths are falling apart, the central myth will not hold. The human revelation is here; its Protestants will stream from the artificial institutions of church, state and city. On 'civilization's' horizon is the redawning of our natural, human, spirit nurtureculture and selflifeworld authentic to us as our birthright, along with these, whatever soul is authentically ours. All these authentic, human elements/spheres are pure as primordial air and water, before they are dragged through the artificial muck of 'civilization' socioculture and religion. It is our choice; to choose continued, regimentation interspersed with irresponsible, adventure, fun and excitement or to choose commitment to recovering yearly and generationally our natural, authentic, human spirit, being, nature, and reality. Because it is either recovery or extinction, some will be up to it.

## 29. Science

The paramount achievement of science to date is the recent realization of the concept and megaphenomenon science calls *biosphere*, or *ecosphere*, which is defined as the entire, ecological system of the Earth and encompasses the water, atmosphere and crust of the Earth. This achievement and understanding on the part of science parallels one of the revelations (or new understandings) of the philosophy of human authenticism, namely, the megaphenomenon or mother-phenomenon it terms Earthlifeworld. Science can now strive to fully realize all the implications and realities of its discovery of the biosphere, which the philosophy of human authenticism forthwith reveals and outlines in this volume.

The central implication for science lies in one major revelation of human authenticism: that Earthlifeworld (in science, the ecosphere, the organismic integration of all ecosystems) includes within it a human social and sociocultural ecosystem revealed in this volume as human selflifeworld. The philosophy of human authenticism reveals that the human realm and the realm of living nature's life are not in any natural conflict as the last three or four centuries of Western, scientific thought would have us believe and as still older ideologies of Western 'civilization' would have us believe. These two realms, the human realm and the realm of living nature are in organismic union when they are in line with their true, authentic natures; that is, so long as humankind—human selflifeworld—remains sufficiently unadulterated and unpolluted by alien, artificial elements-phenomena, and so long as the Earth's natural megaecosystem, commonly called the environment or nature, remains sufficiently unadulterated and unpolluted by alien, artificial elements-phenomena.

Science, then, now has—though it still unrealizes this—the conceptual structures it needs to establish and embrace a general, practical unity of science approximating the theoretical unity of science, for which a few scientists have been calling. All that remains is for the scientific 'communities' and the educational 'communities' to affirm and embrace the realization that the philosophy of human authenticism reveals—the values and human purposes it reveals. And flowing from this revelation in scientific consciousness, science will be able to know and purge itself of those elements and systems of thought and study which are not in line with natural, authentic, human consciousness, perception, thought, purpose and ethics regarding the biosphere which includes the human ecosystem of natural, authentic, human selflifeworld. This resulting cleansed science will be free from destructive, adulterating and alien values, purposes, pursuits and activities, leaving a consolidated and purified science which is generally unified within all its activities. This will be a transformation of science into a new entity, a new endeavor or inquiry which as suggested previously in Chapter I might be termed *philoscience* or *philosence*, indicating that science has been absorbed into philosophy as a supportive organ to the final, *complete*, natural philosophy, authentic to our human subspecies, Homo sapiens sapiens, as it responds to its worsening pathogenesis—sci-technic 'civilization' (alienism).

It has been noted by critics of modern science-technology that there is a long time lag between the rise of problems created by industries from modern, science-technology applications, and the ensuing recognition of and remedial action addressing these problems. That is to say, that these problems have to attain to crises proportions before they are recognized. More troublesome and damaging, the deeper the truth about the human realm and about its relationship to the natural realm, the longer is the wait for science to grasp the truth. Consider science's understanding of the biosphere and its many interrelated systems and ecosystems: the apprehension of this mother truth and reality about Earthlifeworld and its relationship to human selflifeworld—something generally apprehended through diverse, anamistic metaphors, symbolisms and mythologies by all prehistoric societies—has only taken the science of 'civilization' 5,000 years to grudgingly recognize and whiningly declare, with some sectors of science in open revolt against a truth suspected to be implicity heretical to science-technology. The recognition is undeniable now only because modern science and its technology is clearly upon the verge of destroying both the human realm and Earth's natural realm.

The inadequacy and continuing failure of science in its bid as a candidate for knowledge and/or human understanding lies in its ancient birth. The mind penetrated by alien objects or phenomena has no inherent capacity to understand these alien artificials. All of what preceded these alien artificials were givens—as natural parts of selflifeworld to be understood and integrated by the natural growth and perpetuation of natural understanding and intelligence authentic to the species Homo sapiens, just as in previous human species. The unnatural penetration separates the intellect from the rest of the mind and transmutes it. Whole, unified mind thusly fragmented ceases whole, natural understanding; the intellect is assigned the study of the alien artificial: for the penetration also separates ego from whole, unified, triecologic self; and, ego mistakenly accepts the aliens as something which can be understood. Ego, separated and alone is an alien condition resulting from alien, artificial penetration of mind, a condition which does not occur in prepenetration humanity. This alien condition of mind—intellect/egoself—proceeds to progressively build a quasi life out of the study of the progressive influx of alien, artificial objects-phenomena, an influx coming through the punctured membrane or perimeter of species' identity, nature, being, spirit and reality. This influx amasses into 'civilization'; and the above mentioned study becomes philosophy and science.

Science and all its branch sciences must be harnessed because, like the wild horse, dog and the other domesticated animals, science by its very nature (artificial living nature) is incompatible with the realm of natural, human, wholesome socioculture. Science, like a wild animal or alien entity with its own inherent nature, must be domesticated to the purposes of natural, authentic, human socioculture. Changing analogies, science in its living nature can be likened unto groups of various bacteria or viruses for which the human realm of human nature, being, spirit, reality and socioculture has not in its nature evolved any inherent immunity or adaptive capacity. If 'civilization' man must take measures such as vaccines to protect himself from biological aliens—bacteria and viruses—alien to his genetic pools' immunological systems; and, moreover, if he must make laws to protect himself from interpersonal behavior (social, alien phenomena) he judges outside his sense of proper behavior; then surely he must do the same for the appearance and influx of sociocultural aliens—elements-phenomena—for which he has no inherent means to incorporate into his natural, organismic reality, being and socioculture.

If we can make this basic change in what we conceive the nature of science to be, if we can domesticate each of the sciences in the special ways that will make them adhere to and be compatible with the purposes of human being, nature, spirit and socioculture authentic to us, then we can reverse the alienization of human socioculture—the 'civilization' of it—and move in the other direction, into the human reauthenticization of 'civilization' socioculture, i.e., the reauthenticization of this socioculture. This replacement of the 'civilization' of humankind with the reauthenticization of human 'civilization' in its urban areas will regain for us an authentic, human hope for the long term, indefinite preservation of humankind, replacing what continues to prove to be the false hope of sci-technicism. Our hope will be grounded in the real strength of real, human being, nature and spirit, instead of a hope grounded in the illusory power of rationalism, mechanism and open-ended artificiality.

The core value and function of science is that it explains to us the *details* of the alien, artificial conditions or environment that 'civilization' men find themselves beguiled into; while failing, however, to reveal the *nature* of that condition or environment of 'civilization': as the movement out from human nature and being into dehuman, denatured alien being and alien, artificial socioculture.

Science has had no purpose or spirit except the separated specialized 'truth' of its fields. This is analogous to studying one part of the body and then proceeding to practice general medicine. Partial truth inaccurately integrated is still falsehood. Applied science—applied to people, social units and sociocultures—is in violation of any proposed oath for the scientist equivalent to the hypocratic oath, "to do no harm": it is in violation until it at least achieves some unity of science equal to medicine's substantial unity. Or, otherwise until it has a representative, international, policy making congress that restricts and directs the swift, robotic march of the science-technology seen in the twentieth century. National sovereignty becomes damned whereever its development policy of a nation effects unreasonable harm to neighboring peoples and/or to humankind's biosphere.

Where, a good philosopher might ask, is the moral, human guidance for a preventive, human welfare to replace the patch up the pieces and treat the symptoms approach of reactionary science-technology we have found inadequate for so long? And, where is the science-technology not fragmented by nationalism and its own lack of general unity that can care for humankind and the biosphere (Earthlifeworld)?

## 30. 'The Unity of Science' as False Quest

A unity can be either one of integration or organismicity. If science is to have as its purpose the benefit of people and humankind, it will adopt organismicity as its model, with the human sciences as its major, vital organs and other sciences as minor essential organs. For its heart this 'organism' needs anthropology. For its mind it needs a new philosoence, which pursues the good for humans and humankind, which is, of course, already contested within philosophy by such contestants as ecology, ethics, social philosophy, metaphysics, anthropological philosophy, epistemology and logic. Since what is good for man depends on what the complete nature of man is, a complete definition and/or description is called for. This new inquiry, if intact, unadulterated, monistic and resistant to abnormally swift sociocultural change, could also handle philosophical anthropology and anthropology.

A unity of science entails within reach an objective, collective consciousness of a collective nature of and goodness for man and a collective self-consciousness that will embrace it, and the self-consciousness of it as its own, a priori or through revelation. Such a revelatory, a priori, archetypal, natural, authentic human condition for our 50,000 year old subspecies, Homo sapiens sapiens, does exist as our conscious human being, its nature, spirit, reality and selflifeworld that is authentic to it. It is revealed and presented by its philosophy, the philosophy of human authenticism. It is the 'organ' of consciousness for the organismic unity of the human and natural sciences, and human philosophies.

Natural science by itself is utterly discredited by the revelation of human authenticism, namely, the revelation that natural science has all along been beholding living nature with an unnatural consciousness, mind, intelligence, methodology and perception. There is no way that living nature can be truthfully or accurately beheld, perceived, understood or explained through unnatural, alien, artificial consciousness, mind, intelligence (intellect) and perception.

The dream of unity, when in human affairs, is a dream of human life. For, unity in human and all life is organismicity; whereas in inanimate matter, physical science, mathematics and mechanical 'civilization', material and conceptual unity is merely mechanical and rational integration. Thus, the more we can organismicize human relations, social units and societies, the more natural, real and meaningful they become. Otherwise, the more we achieve of mere integration in human affairs and socioculture, the more we become like integrations of material objects and ideas—like cybernetical machines.

Ultimately, science can only strive for an improved unity toward an optimum unity. It can never achieve a genuine, complete unity of science; for, it is the essence of science to tear things apart, to pull the objects-phenomena, cognition and consciousness of human selflifeworld apart and then to reorganize them back together again utilizing a transmuted, quasi intelligence—intellect—, a transmuted, quasi perception, mind, being and human reality, all of which belong to the quasi intellect-egoself, its quasi life and quasi world. It can achieve a unity only of a conceptual, mechanistic integration. Genuine real unity belongs to living nature, part of which science rejects or at times subordinates: it rejects natural vision and senses, natural perception, natural intelligence, natural consciousness, natural mind, natural being, natural reality, and their natural selflifeworld.

# 31. Confucianism and Taoism

Confucius embraced and defended the Way (Tao) of the ancients just at the time when this Way had lost its ability to regulate social and political discourse. He did so with a new interpretation of the Way which proposed a moral justification for the old hierarchical society with the presupposition that this old society corresponded to a natural moral order. This presupposition is clearly false in the light of recent anthropology revealing natural, pre-'civilization' socioculture and a'civilization' Earth-folk socioculture to be largely free of formal government and social hierarchy. The onus was, thus, upon each individual to assume the moral obligations inherent in his position as son, father, subject, ruler, and so on. The cardinal virtue, jen, of the Confucian morality, in fact, embraces all the Confucian moral qualities of the true man: loyalty, reciprocity, dutifulness, filial and fraternal affection, courtesy, good faith and friendship.

Confucius, therein, for all his adaptive, beneficial teachings, had fallen victim to the original and still classic error of philosophy, i.e., the fall into the pragmatism of accommodating the urban state and its departures from natural, human order and laws of human nature and being. Further still, Confucius fell victim to the generating mechanism (the dialectic) of the urban state and city-state—the power-wealth spirit-ethos. He sold out for position—for power-wealth. Devoured, like nearly all those trapped in 'civilization' sociocultures, by the two-headed monster lurking in the intellect-egoself, i.e., statism and power-wealthism!

For a time, Confucius traveled from state to state in an unsuccessful attempt to win the support of their rulers for his Way. But, although some of his followers eventually gained high office, he regarded himself as having failed in his mission. The reason for his failure was too drastically simple for even Confucius to apprehend. Confucius' embrace and defense of the Way was predominantly of the branch called Lao-tzu which unlike the branch of Taoism, Chuang-tzu, is conducive to attracting and attaching the state as a sociocultural condition judged compatible with a natural, moral order—natural, moral, human order. Confucius had not experienced the human revelation of human authenticism which reveals the state as a recent, alien, artificial quasi human socioculture opposed to and therefore incompatible with a primordial, natural, authentic, human socioculture 50,000 years old that preceded the state and that fostered the growth and perpetuation of a natural, human, moral order as well as a natural, human nature, being, spirit and reality authentic to our natural human subspecies.

The emergence of philosophical Chaung-tzu Taoism quite likely represents the first, great, clear-cut, open sociocultural struggle within a 'civilization' between the dialectic of human authenticism and the dialectic of alienism. Chaung-tzu Taoism emerged as an early, systematic philosophy in a dramatic, strong opposition to 'civilization' and its obscurely defined, unarticulated, unsystematized philosophy of alienism. This philosophical Taoism is probably the most legitimate precursor philosophy to human authenticism; emerging over 2,000 years before the West's postindustrial revolution finally pushed Western consciousness against the wall and forced the dawning of a more extensive human revelation, the philosophy of human authenticism.

This original Chaung-tzu, philosophical Taoism, alas, fell short of the complete human revelation; it was an early, oriental romanticism, recapturing primordial essences of natural, human living or selflifeworld. But, the impetus toward a genuine, authentic, human freedom, a liberation of the human condition from the restrictions of artificiality's adulterations, became stalled in the valley of individualism. This Taoism held that for one to be free and virtuous is to be free from the bondage of circumstance, personal attachments, traditions, and the need to reform the world. While this is practical for the philosopher and the solitary, meditative, single person, to be fully human and fully free is to both have unity with one's human nature, along with having an instinctive unity with living nature, or the Way. As stated in "Freedom and Determinism," in Ch.X, authenticism's revelation of a natural, species nurtureculture, intergrown with biologic, human nature, establishes human nature as being wholly natural. Freedom for any natural animal being and human-animal being can only be the condition of being in line with one's species nature. For, we humans, as social animals, human nature (and being) includes more than an individual, an intellect-egoself, moving out of restricting artificiality of 'civilization' socioculture; it includes the rest of the whole, natural, human, triecologic self, i.e., family-self, and community-self. This natural, human (and social animal) self, authentic to our human nature and being, includes a natural family and a natural community, and, moreover, includes a natural lifeworld that is to be lived continuously rather than to be made contact with at intervals, or to solely maintain continuous contact with living nature (hermit-like) thereby avoiding artificialized family, community and their quasi life and quasi world.

Every romanticism fails to come to grips with the origin and nature of artificiality: they fail to account for how and why circumstance, personal attachments, traditions and activities come to be bondages for us. They fail to have the revelation that for the individual to commune with living nature (in Western expression) or to follow the Way, the Tao or Buddha nature (in Eastern expressions) is to free only the intellect-egoself, i.e., a part of the whole self, from its bondage to artificial circumstance, social relations and socioculture. This leaves it exiled from the naturals of these, from portions of the whole, human self and from the portions of human life, of human nature, being, spirit and reality that are indwellant with a natural, nurtureculture socioculture which preceded and was adulterated—artificialized and alienized—by 'civilization' socioculture with its adulterated, surrogated, diminished, fragmented, transmuted circumstance, social relations, traditions, activities, perceptions, conceptions and consciousness. In short, all romanticisms and the isms springing from or reacting to them falsely entail the fallen intellect-egoself and its transmuted quasi life and quasi world as the natural basics of the human condition. Human authenticism reveals the original, natural, real human condition, authentic and fitting to us by revealing the existence of natural 'culture'—nurtureculture—devoid of all or of any significant transmuting artificiality, and revealing the existence within this purely natural and human socioculture (triecologic selflifeworld) of natural, human consciousness, mind, being, reality, nature, and spirit.

It is one thing to merely open urban consciousness to a rich awareness or 'sensing' of living nature's megaorganismicity; it is still another to open the way for human being, nature, and spirit to play their evolved, ecological, natural part in that greater natural ecosystem. Western transcendentalisms and Eastern mysticisms intuitively or instinctively grasp, and Chaung-tzu Taoism rationally describes a consciousness of what is largely hidden from urban, 'civilization', socioculture consciousness, i.e., the manifold 'sensings' to be discovered in living nature that the individual's consciousness can experience moments or periods of union with. However, even gathered in groups these individual consciousnesses, moving through the Way into the experience of partial, periodic unity, are still blocked from the way to experience full, natural experience and consciousness of full, natural, human life (selflifeworld), being, nature, spirit, and reality. Only one revelation can preserve us now, the revelation that living nature—social, animal, primate nature—has evolved for us a natural, human-animal 'culture', i.e., a natural socioculture—nurtureculture—that is intergrown and interfunctional with our natural, biological nature, the revelation of this natural, human selflifeworld that is within natural Earthlifeworld, that is by living nature's and human nature's ecosystems, our authentic, *human* Way. At this late hour this is the only 'way', the necessary way to humankind's preservation.

At bottom, it seems that several major isms have pursued some sort of unity with the laws or principles of living nature or of a Creator of living nature. They have, in this volume's view, sought some closer relationship, harmony or instinctive unity with the megaorganismicity, Earthlifeworld. The absence of any complete philosophy with a full, systematic explanation for how and why humankind and/or a human being comes to be outside this organismic unity so variously described by diverse sociocultures can be explained be the absence of two recently arisen preconditions for an unprecedented human revelation to 'civilization' consciousness. These have arisen simultaneously, as two interlinked, preparatory, growing subrevelations. One, that science and technology, and the artificialization that they are the general manifestation of, have moved well beyond what is generally beneficial into a general destructivity that portends human extinction. And, two, that progressive sci-technic society has been falsely perceived as the inevitable, the right and the only sociocultural model; while, to the contrary, another destiny is increasingly recognized as available and necessary for humankind's preservation, one based on the values and models found in pre'civilization' and a'civilization' Earthfolk sociocultures. In short, a growing awareness of the intrinsic, decayous and self-destructive nature of 'civilization' is occurring together with a growing awareness that simple, more natural socioculture lacks this overt—and recently, progressive and excelerating—destructivity.

## 32. Individualism

Individualism emerges when the community-self and family-self are penetrated and fragmented, start breaking away in pieces from the bio-egoself. When the community-self and the family-self portions of the whole self are sufficiently diminished, a point is reached where the remaining bio-egoself must make some kind of cohesive perception or 'sense' out of itself as bio-self—a fragment of whole self—and the remaining fragments of life and world; it utilizes intellect for studies toward this end. It bonds to the intellect portion of mind and intelligence, becoming transmutant intelligence. It requires new doctrines, values, and thought systems for its philosophy of living, as well as, new organizations for its quasi life and quasi world. (*See also* Ch. IV, Part B.)

The individualist dwells almost wholly in the personal bio-intellect-egoself and its quasi life and quasi world, and posits this egoself as the dwelling of human being. He moves through life with its mortality at center of consciousness, instead of his immortal elements of human being and spirit at center, and mistaking bio-intellect-egoself for whole, human self. Finding this egoself fallen and deficient in human being, he commences a career of

search, or adventure-discovery, and exploration for a more full experience of human being. Thus, he and all others that constitute
the multitudes within 'civilization' consciousness, begin an inef-
fectual struggle to recover part of their lost human being.

For some, recovering and experiencing our human being becomes an intellectual career or a lifelong spiritual struggle. Such
individualists are the wisest of the fallen. They know that an essential portion of the human self and of humankind's collective self is
lost and is separated from missing portions of lifeworld; and, that
these losses and separations need to be recovered and reunified,
respectively. They have erred, however, by placing human being in
a erroneous dwelling—in the biologic, intellect-egoself and its quasi
life and quasi world. Natural, authentic, human being does not
dwell there; it appears there only in visits to the fallen egoself,
imprisoned in an alien, disintegrated quasi self, quasi life and quasi
world. Human authenticism frees the prisoner with the most potent vision available to 'civilization' man, the revelation of what,
who and why he really is: the vision of natural, authentic, human
selflifeworld.

An individualism, so deeply advanced into modern, sci-technic 'civilization' selflifeworld—so deeply lost—can only attempt
and succeed at a sense of human being and brief experiences of it
in its original, authentic fullness. Authenticism delivers it up as
the rebirth of natural, authentic, human nature, being, spirit and
reality with the recovery of the selflifeworld of these.

We have placed a large part of our good faith, through
individualism, in the ego. And, it has progressively turned against
us, effecting our social, moral, spiritual and sociocultural
disintegration through the masquerade of an expanding,
materialistic, rationalist, hedonistic, anarchic, mechanical creativity
that leaks through its widening cracks much of our human essences,
i.e., much of our human spirit, being, nature, and selflifeworld. It
is a sham creativity, an egoistic, intellectual and mechanical one
that is our leader and leads us ever deeper into a disguised
destructivity, dressed in sci-technic novelty, exploration, innovation,

fascination and fantasy. This egoistic creativity and the individualism it is a part of emerged from two central origins. One, it emerges from a mere part of the whole, human self, the ego, which is too incomplete and immature to produce or create a fully human truth, beauty, goodness and reality. (It is no coincidence that the emulation of youth to the point of becoming a sociocultural, youth ethic reaches its highest manifestation in the USA, the socioculture that has pushed the farthest into individualism. Nor is it a coincidence that only Euro-American history records a famous and admired quest by explorers for a fabled fountain of youth precisely during the early surge of Western individualism.) Two, this sham creativity emerges from the city and state of 'civilization' socioculture grown too large to recognize, respect and uphold the authentic, real needs of the whole, human self, its organic family and community portions, and its natural lifeworld authentic to it. Real, natural creativity of any real, natural sociobiologic being springs from and is directed from within by its natural evolved human being within, and as well, by the natural ecosystem of its natural, sociocultural units and natural lifeworld permeating it and around it. It is the creativity of natural, organismic growth and perpetuation. It is not the coping, adaptational development of ideas, gadgets and systems that, while growing progressively more large and complex, progressively lose the natural, harmonious, organismic function of living nature and human nature and being during the vaunted process, of growing outwardly in their mechanical, material, conceptual, organizational and artistic accumulations.

There is no such entity as an individual human being in precise truth—but only in convenient perception and conception: there are individual, biologic beings; but, human beings are intergrown and interfunctional as social units of family and community through the social part of their sociobiologic nature and being. Our biological, material bodies are individual things/parts; while our social/spiritual, nonmaterial beings are intergrown by the very living nature of social/spiritual beingness. It is a contradiction to think of and an illusion to believe in individual human being, since to have human beingness—to be a human being—one must grow that bio-being, intergrown and interfunctional, with other beings within natural, family being, natural, community being, lifeworld being and natural, environmental (Earthlifeworld) being. A unique body, mind and personality are parts that together form an organismic pact or unit of human being that is intergrown and interfunctional with the family, community, lifeworld and natural environment to constitute whole, human being. We are aware of a human being only when our natural bio-ego consciousness or our 'civilization' intellect-egoself consciousness focuses on the unique *part* or portion of human being which in the natural consciousness of natural, authentic, human selflifeworld is merely secondary, subordinated consciousness to primary, general consciousness of whole, human being, which is structured/functioning/dwelling in human selflifeworld.

## 33. Language, Reality, and Being

That portion of human reality that is intended or evolved, ecologically to be communicated from person to person and people to people through the element or faculty of language cannot any longer be communicated completely once humans have come to live within the lifeworld of 'civilization' socioculture. 'Civilization' socioculture is artificialized by definition—adulterated, fragmented, diminished and surrogated—therefore, all that is spoken or written about the natural, authentic human realm or human condition is no longer fully realizable or representable as language: all these are now partials, along with the main spheres of adulterated, artificialized 'civilization' lifeworld, e.g., partial reality, partial being, partial communication and partial language. This is not because they are reduced in quantity, but because even though they have increased in quantity, the additions have been adulterants and surrogates which have decreased the degree of genuineness or realness of human reality, human being, human communication and human language. Their quantitative increase was not totally natural, genuine and real; instead, the additional influxes were human pollutants or adulterants. Thus, what we perceive to be growth is actually polluted, infected bloating or swelling, since the pathogenesis of the human 'civilization' realm is generalized, encompassing the material, physical, mental, moral and spiritual. 'Civilization' is, therefore, a gradual, evolving process of human, sociocultural insanity. We are losing our personal, social and sociocultural health and sanity just as surely as a person infected with the syphilis bacteria gradually loses his health and sanity.

## 34. Sight and Language

We regard humans to be visual animals in general human affairs. That is to say, in moving about, locating food, using tools and utensils, and other basic activities, it is sight that is used often alone or in essential eye/hand coordination. However, in our time language has been credited an increased role in the constitutional nature of the human condition, even to the point of being regarded by most as the distinguishing feature of our human species. Still, language does, after all, involve the sense of hearing, sight of facial and body gestures, and the sense of touch within the mouth. The modern emphasis upon language together with these involvements of hearing and touch in the phenomenon of language, challenges the previous notion of humans as markedly visual animals. Challenging the centrality of either language or vision is recently and continually arriving knowledge of pre'civilization' sociocultures and a'civilization' Earth-folk sociocultures, including the present and past prehistoric ones that have brought to light the importance of other human elements such as dance, drum beats, ceremony, custom, ritual, folklore and mythology. At present, a more correct understanding of human nature, being, spirit and reality, and their natural socioculture affirms a more balanced, organismic orchestration of the senses and other mental faculties than we, in our modern biases, have heretofore realized.

Still, some may persist in believing that sight and language are two mental faculties that have an edge over the others. Regardless of the truth or falsehood of this idea, more importantly, these two have proved to be the most susceptible to dysfunctional inflation and transmutation by 'civilization' sociocultures. Other imbalancing excesses attaining to human transmutation, of course, naturally abound in this volume concerning human transmutation. To be sure, the particular details of human transmutation away from natural, authentic, human socioculture into 'civilization' socioculture and, finally, into sci-technic 'civilization' are, in themselves, very interesting. But, the primary importance here lies in understanding why these transmutations have occurred and what they mean. What is the essential meaning to us and our progeny of this human transmutation, long-running and gradual, involving at first the tiny minority of humans living in the first cities and, finally contemporarily, about fifty percent of humankind? What does it mean to us as denatured, natural, living, dehumanized, human beings having experienced a substantial measure of artificialization and alienization of our human being? It either means what 'civilization's and their histories (including the reader's) say they mean: human progress or betterment and the development of desirable human potentials previously unrealized; or it means that the natural, human nature, being, spirit, identity, reality and human selflifeworld authentic, right, fitting and belonging to us is gradually being transmuted through adulteration, fragmentation, surrogation, diminishment and disintegration into an artificialized, alienized, decayous human condition. It either means that we are experiencing the evolutionary growth-fulfillment, perpetuation and slow change allowed within living nature's and human nature's laws; or it means that we have departed from the slow, adaptive evolution that living nature grants our species according to its nature and the nature of its relatings to living nature's megaorganismicity of diverse, living natures, and, therefore, are diverging deeper and deeper into a nonevolutionary, nongrowth, nonperpetuation, that is to say, into a progressive decay leading, if not reversed, into an extinction or decimation of our human species.

# 35. Language

In recent decades linguistic philosophy, philosophy of language, linguistics and sci-technic society in general have largely identified language as the central human phenomenon. This error stems from some explosions that have occurred within this unprecedented type of socioculture. The explosive growth of science and technology resulted in the explosions of scientific knowledge and data and of technology's techniques, both of which immediately have attached to them the terminologies or language required for these two interconnected activities (interconnected from Galileo on) to grow explosively. The desire of these two interconnected "communities", the scientific and the technological, and others to be heard resulted in another explosion, i.e., the communications and information explosion. This is the filling of an artificial need, desire—of exploding science and technology to be heard,expand, and gain power through the creation of a new science and technology of communications.

These explosions together with the explosion in the variety of the occupations available and leisure time activities available, all together can be correctly understood as an explosion of socioculture itself, i.e., as explosions occurring within the larger explosion phenomenon of sci-technic socioculture's adulteration, fragmentation, surrogation and transmutation of human socioculture's constitutional nature, authentic and natural to it. This explosion of socioculture is an explosion of sociocultural objects-phenomena wherein kinds and shear amounts of objects-phenomena are multiplied. Simultaneous with this sociocultural explosion and the communications explosion within it, there is, then, an explosion of language. Language, then, resultingly has a greater task thrust upon it than what is intended in the evolved nature and the natural purposes of natural, human language and in the ecological design and natural purposes of human being, nature, spirit and reality that are authentic to our human subspecies. Language, then, is falsely perceived as the problem with the supposed solution lying in the successful study and an improved understanding of language. Correctly perceived, the problem of exploding language springs from the more basic and real problem of exploding society—sci-technic socioculture—exploding beyond the boundaries of the natural socioculture that is authentic to us and our human nature, being, spirit and reality.

Meanwhile, language undergoes a different kind of deterioration aside from its explosion. It receives a mutilating assault within the sci-technic countries of 'civilization' socioculture. It has become an expedient and coping belief on the part of growing numbers of sci-technic peoples that one can change the very definition and facts of what the phenomenon of language is simply by using it in different ways and applying the term *language* to new phenomena making these additions inclusive within the popular preoccupation with language. This that befalls language has been witnessed before when art was struck by the phenomenon of modern art, specifically the idea that holds that what is art is purely a subjective evaluation, that almost anything created by almost anyone attempting to produce art can be legitimately labeled art so long as the artist and at least one critic or credentialed artist, or one reporter needing a story, agree that it is art. Indeed, the loose and dubious criteria for what is modern art closely resembles the criteria for what is modern sanity. Thus, in the same manner by which sci-technic society's new, subjective, ethnocentric criteria are devised for what is to be regarded as human reality and human being; so, too, language is to be adulterated into redefinition and transmutation from its authentic nature and function for human being and society. It is language's turn to undergo misguided surgery upon its nature as human phenomenon.

In our era when language is held to be of paramount importance in understanding human being and reality by science and philosophy, the result is that the scientific study of almost every other form of human communication—such as body language and sign language—scramble for a share of this attention and glory by proclaiming themselves inclusive within the language phenomenon rather than related but distinct forms of human communication. Thus, while for thousands of years—long before science and 'civilization' emerged—people have known that language is voice sounds intuitively and instinctively created into what has been regarded by common (human species') sense as language; now, we are confronted by obsessed, specialized experts telling us that we humans have always incorrectly understood what language is, along with everything else about ourselves and our lifeworld! And, telling us that language is now to be redefined to meet the recognition needs of these specialized, individual and collective intellect-egoselves. So, if we cannot hold language together to do what its natural, authentic, human function is and cannot halt and reverse its mutilation, through excitation based novelty and experimentation, by the scientific sector, the mass communications and entertainment industries, and the diverse pop subcultures, then each generation merely takes what it finds remaining in the way of exploded, fragmented, mutilated, adulterated and transmuted language and communication, and thereupon redefines it as language. Thus, it is to be that language is to be a crucial tool in the progressive deterioration of human reality and human being (progressive insanity).

The whyfore of all this: that the doctrines, ethics and ethos of the egoself—the individual intellect-egoself and the various collective egoselves—must be upheld at all costs because they constitute the transmuted, quasi, egoself will that is engaged in dialectical, mortal combat with evolutionary, natural, human will, or purpose, authentic and fitting to natural, human species' nature, being, spirit, and reality.

It may very well be that the task of the science of linguistics is to react to this new problem of exploding language commensurate with its understanding of the problem. However, the task of philosophy is not that of reaction *per se*, and in this case, not of reacting to what is happening with language, but to understand that what is happening with exploding language is the result of socioculture exploding; and, then, to set about analyzing and judging whether this *should* be occurring, and if not, what are the dynamics behind this deleterious occurrence and what can be done to stop or decrease the phenomenon; the task is not merely to react to what *is* occurring. The philosophy of language and linguistic philosophy have failed and will continue to fail because they focus on a resulting phenomenon rather than on the primary, base phenomenon which is the occurrence of exploding socioculture for which the task of philosophy is to define the nature of what is happening and determine or judge whether it should be happening and, if not, to propose remedies or alternatives. More succinctly put, philosophy should have the wisdom to distinguish what is from what should be and to propose the ways in which what is happening can be brought, as much as possible, into line with what should be happening.

Many are led to worry about the threat of multiple, nuclear explosions of matter. After all, the philosophy and consciousness of materialism—that matter is the central reality of perception—continue to reign. Meanwhile, the largely, non-material explosion of sci-technic 'civilization' socioculture continues, with its disintegration of the social, mental, spiritual and moral realms, and the natural environment.

# 36. Television

Television surpasses its electronic, artificial relatives in presenting artificial sight and sound, and goes further into the mutilation of mind, being and reality in its presented image of directed, imagined and occurred or occurring scenes or news events by means of fixed, artificial lens or lenses (cameras) of fixed, selected, timed, aimed angles. Television is the democratically sanctioned mutilation of humanity by the union of commercialism and sci-tecnicism. It mutilates the senses of the viewer, the program presented, and vicarious experience itself. It mutilates the mind through the above and further by imprisoning the eye and ear to the TV set, to the director's eye and ear, to the writer's mind, to the camera's view, focus and angle; by imposing the director's and/ or writer's interpretations of consciousness, meaning and purpose upon the mind, heart, character, and personality of the viewer. Television, through an hypnotic daze of excitement, novelty, amusement, fascination—through entertainment—with casual aplomb, destructively ransacks the manifold elements of natural mind, life, reality and being of the viewer. It removes the viewer from both real, vicarious experience and real experience of self, family, home, community, reflection, absorption, work, social activity, social milieu, natural milieu and the other elements of human selflifeworld. It requires from the viewer an analysis (sometimes long and deep) of the director's and/or writer's interpretations and a formulation of the viewers own interpretation whenever these two are significantly different; or, as more often happens, the artificial experience of TV results in passive acceptance or indifference on the part of the viewer regarding these given interpretations.

Television achieves a quantum leap in fragmentation of human mind, intelligence, being and reality. It separates eye and ear from natural, authentic mind and self, and from objects-phenomena of natural, authentic lifeworld; it separates egoself from family (what's left of it) and from community; it separates family from community (if any); it separates mind and body from nature, being, reality, and spirit; it separates mind from spirit; it separates time from being, nature, and spirit.

Television can only be experienced by the egoself. A person— a whole, triecologic, human self—is mutilated by the watching of television leaving only the egoself. A family all watching the same television show does not escape the mutilation of TV. All of the mutilations, impositions, separations of this alien phenomenon which reduce whole, human self to egoself cannot be escaped by a family banding together for family viewing. They are overpowered simply because viewing TV is an alien mutilating experience. All alien experience is unable to nourish or sustain natural being and spirit nor participate in natural, human mind and reality. Viewing TV is an anomalous act of self-mutilation; a temporary loss of sanity. And, it must by definition damage natural being, nature, and reality just as artificial elements or compounds in water, air or tissue reap organismic damage. Living organisms have evolved no defense against artificial, alien objects-phenomena. Humans do sometimes *devise* alleviations and bandages. These are not defenses and do not prevent or reverse generalized damage.

With television, the heart of the message is the medium, and a medium that is also a mission: to remove people from participation in natural, real life—in their selflifeworld; to have them observe, and sensually, emotionally and conceptually process a salad or soup of data, impressions, images, depicted emotions, thoughts and phenomena, rather than amass life experience—experience human being.

Television is a toxic, goulash soup for the mind, containing syntheticized vegetables and artificial seasonings, flowing into human consciousness; the result being a diarrhea of consciousness—undernourished and polluted thought, emotion, perception, intelligence, reality, spirit, being, family, community and lifeworld.

Television can never be a part of one's real lifeworld. Changing metaphors, it is an alien, artificial monster of pleasure entering not through our bloodstream, as do sweet, alien chemicals do, but through a continuing, audio-visual explosion that sends electronic shrapnel tearing through our eyes and ears, tearing through our minds, personalities, characters, emotional structures, idea systems, value systems, belief systems, behavioral patterns, relationships, families and communities: hypnotic, pleasurable escapism into the mutilation of our human nature, spirit, heart, soul, being and reality.

# 37. Photography

Photography does to our visual images what the written word does to our spoken word. The written word was the great transmutation of the spoken word. Photography is the great transmutation of our visual images, the transmutation of what we see. It largely removes from vision any depth, relation, context and movement; it reduces the role that memory, imagination, and intelligence play in our past and present; this in much the same way that the written word took a great deal away from oral communication, oral history, the storyteller, mythology, legend, ceremony, dance, ritual and custom.

The motion picture adds some movement, relation and context, but for elitist purposes and values. The mind, instead of picturing with memory, imagination, and intelligence, according to its needs for identity, growth, purpose and being, is presented with a storm of non-prioritized pictures, still and moving pictures, according to the elitist purposes of the photographer, director, writer, producer, advertiser, et al. When the desires of elites for artistic expression, fortune, and fame override the people's natural, authentic, human needs for growth, identity, purpose and being, therein we have yet another tyranny and oppression, i.e., that of the commercial synthesis of fortune, fame, talent, art of creative destructivity spread like artificial garden plants upon the path of humankind. The people's authentic human needs are ingeniously replaced by what the people have been conditioned and programmed to want and desire.

Photography's only vindication is in its art. The photographic artist's only justification is found in recapturing parts of what has been lost of natural, authentic, human being, nature, and spirit. He is not to create new, adulterant artificials or rearrangements of adulterants. The true artist is not an entertainer nor even a mere teacher; he must be a healer: help human being recover some of its lost growth and fulfillment. Art must nourish human lives and souls, not spark interest in novel, adulterating, artificializing compositions about the natural realities of the human and natural realms.

# 38. Childhood

The dilemma of unnatural, unauthentic childhood can be conspiciously seen in the problem of unauthentic, unnatural toys. Toys and playthings in general display a progressing phoniness in sci-technic socioculture. Witness the child in the toy department of the store. He is presented with a myriad of objects represented as 'toys'. These toys have digressed from the play objects of human nurtureculture, which are to satisfy the functions of original, natural play objects of Earth-folk (a'civilization') socioculture, which engaged and activated one or more elements of the child's growing, authentic, identity, spirit, being, and reality. The modern child views and examines modern toys with the intention of actualizing and nourishing some part of his growing nature and being. After arriving home with his selection, he often times finds that the object does not, after all, maintain interest into an engagement or that the engagement was strained and ambiguous rather than free flowing and naturally suiting. These toys that the child finds 'does not suit him' arise from at least two basic causes. First, the child was caused to select them through the influence or forces of fadism, commercialism, novelty, the misrepresented toy and malrepresentative advertising—his own innate judgment was swayed by other judgments; his natural toys have been displaced, his need for them suppressed or nearly destroyed. Second, the child in sci-technic society for the first time in history, substantially speaking, has instilled within him phony values; and, more important, emanating from these, are phony, artificial elements of personality, character, identity, self, and lifeworld. He is already at a tender age partly unhuman and partly unreal: he is diverging, through the forces of unauthentic nurtureculture or socioculture—of 'culture'—away from authentic, human nurtureculture and nurturome. In other words, even if false, unnatural toys of un-

natural, unauthentic nurtureculture—'culture'—were removed
from the home and lifeworld, his lifeworld would still retain other
alien artificials, his liberation would be partial.

The truth of the matter is that before 'toys' were invented by
'civilization', the original play objects of childhood were all about
the child in camp and village. Every utensil, tool and every other
lifeworld object are objects of wonder, interest, puzzlement and
experiment until the child, item by item, takes the steps emulat-
ing the use and function of each in selflifeworld. By puberty all his
toys have become his lifeworld objects. He now employs them in
daily life and creates them complete—the ladles, dippers, bowls,
baskets, footwear, clothing, bows and arrows, arrowheads and other
utensils and tools. In natural, authentic, human socioculture everyone is creative craftsmen diversified; and unified, authentic mind
continually recreates lifeworld and self in the maintenance and
perpetuation of natural, authentic, human selflifeworld. The presence of 'toys', play objects that do not lead to use, function and
creative growth/perpetuation of authentic, human selflifeworld is
a mark of 'civilization', of the decline of human species' selflifeworld
and its human being, identity, nature, and reality.

This is part of the original opposition of man to nature, i.e., it
being unnatural for children to play with (grow towards) anything
that is not of, or alien to, their family-community selflifeworld.
And, it is unnatural for parents to permit or encourage them to do
so. In both natural, social animal life and natural, authentic human
life, every moment is a moment of growth, a moment of stagnation
or a moment of decay. Hence, the timeless directive from parent to
child: "Stop wasting time. Here . . . " and the child is given an
object or activity of growth. In natural, authentic adulthood
drudgery does not exist, for in the nurturing of children, the
guidance of them, there is the fulfilling reproduction of a gratifying
version of ones self and ones family. This, along with the creative
growth/perpetuation of lifeworld is the second portion of human
reproduction, perpetuation and immortality. When children are
grown they are grandchildren, as well as, the recreated subjects/

objects within human, phenomenal selflifeworld. Thus, one of childhood's first transgressions from natural, authentic human nurtureculture is illustrated dramatically by unnatural, inauthentic 'toys' in their open, flagrant destructive assault on behalf of 'culture's' vested, artificial, commercial and institutional interests.

However, a second basic category of childhood, nurtureculture divergence is the earlier violation through culture of genetic human nature and being. It is comparatively less disturbing and deleterious (at present) compared to the first category—the violation of human, nurtureculture, nature, and being—which we are addressing in this volume. Suffice it to say here that this less important transgression consists mostly in two actions. First, the acceptance into the genetic pool (local or societal) of genetically grounded features and behaviors that significantly violate, by qualitative or quantitative effect, the authentic nature of human selflifeworld—triecologic self unified with nurtureculture's lifeworld. This is most exemplified in the acceptance of infants into the family and community that are so defective that they represent a deleterious force, which through burden or imbalance, outweighs the beneficial effect.

Second, the 'cultural' violation of genetic nature and being is occurring through the acceptance into the ongoing genetic pool of—the reproduction of—genetic features and behavior that significantly violate the natural, authentic nature of human socioculture's selflifeworld. An example being the acceptance of the practice, with or without community, tribal or governmental policy, sanction or law, of the conception of and carrying to birth—the creation of—defective babies that by the nondefinitive trade-off evaluation of genetic anomalies given above, would be on balance destructive or deleterious to person, family and community (and selflifeworld).

Returning to the first category and the focus of this volume, i.e., socioculture's violation by 'civilization' of our natural, authentic sociocultural (nurturecultural) nature authentic to us, there are two other types, more disturbing and outrageous than toys.

First, the displacement of natural childbirth by the hospital delivery system of sci-technic socioculture. Natural, human intuition and intelligence have, starting in the 1960s, formed an increasing opposition to this mutilation/surgery upon fetus, baby, mother, mother/baby needs and bonding: an outset mutilation/adulteration of the human nature, being, spirit and reality, natural and authentic to us. The reader may turn to a host of authorities on natural childbirth and childhood itself, publishing steadily through the last two decades (e.g., Pearce, Joseph C. *Magical Child*. New York: Bantam Books, Inc., 1980, the Notes on Chapters 6 and 7 and Bibliography.).

To give a more detailed and enumerated account of the violation of natural, authentic infancy and childhood would be to depart from the scope and purpose of this volume. However, herein, the forceful and angry charge is that the transmutation of childhood's whole selflifeworld into a fragmented, diminished, surrogated, adulterated bio-egoself, quasilife and quasiworld is quite far along and gradually progressing against the opposition, part of which is mentioned above. The old dispute of whether 'civilization' sociocultures and any particular 'civilization' collapse more from outer assaults or more from inner deteriorations is herein discredited. These two are not merely happening simultaneously as one would first conclude; but are now revealed as two parts of one phenomenon, i.e., the penetration, adulteration, fragmentation, diminishment and surrogation of human being, nature, reality, spirit, and socioculture, natural and authentic to 50,000 year old Homo sapiens sapiens. The inner deterioration is moving deeper within; the outer assaults are now being made upon First World 'civilization' sociocultures not from each other as much as from themselves. The pathogenesis goes inward and outward.

This recently arisen (comparative to our subspecies' 50,000 life span) struggle of 'civilization' sociocultures—alienism—against human authenticism, however, now is at its final turning point: 'civilization' man is being struck by his final, complete, redeeming, human revelation—of human authenticism.

# 39. Contemporary Philosophy

Toward realizing the state of contemporary philosophy, it will be useful here to refer to the concluding chapter in Ayer, A. J. *Philosophy in the Twentieth Century*. New York: Random House, 1982. Here, Ayer posed a question to himself and the reader, after concluding that modern philosophy has failed to be of significant or substantial benefit to contemporary man: "Why, then, should one even bother to continue with philosophy?" And his answer was: if for no other reason that all the other philosophers are continuing to do what philosophers do—continuing with philosophy as usual. In other words, however unsatisfactory the state of philosophy may be, philosophers have nothing else to do but to continue what they are doing in philosophy. This pathetic attempt to justify what we call modern philosophy and contemporary philosophy, of course, is unsatisfactory and fails as philosophy's justification. Ayer, in a way, is falling back on the words of the father of Western philosophy, Socrates: "The unexamined life is not worth living". This thereby implies that the worthwhile or justified way of living is to examine life. Surely, the justification of life or the judgment that life is worth living is certainly acheived on the more simple grounds that everyone else is continuing with life and life's activities. However, this attempt to transfer that justification for living over to the justification of a formal institutional activity of 'civilization' lifeway, specifically to philosophy, is not successful or valid.

Generally speaking, human life has needed and needs no justification up until that point is reached where a conspicious number of people find life so miserable that they are committing suicide, being self-destructive, falling into despair, or failing to reproduce an adequate regeneration. The idea that life is worthwhile up to the point that it is felt to be unworthwhile or unbearable cannot be applied to particular activities of particular sociocultures such as philosophy. For, human life and life itself have their own inherent purposes and justifications. Philosophy, on the other hand, has only one legitimate purpose and justification, i.e., to have on balance a beneficial effect upon human being, nature, reality, and life (selflifeworld) and its socioculture. Thus, philosophy cannot justify itself with the reply, "Philosophy goes on.", the way that life can justify itself through the fact that "Life goes on."

Personally, the individual philosopher can justify his own pursuit of philosophy through the fact that it benefits him; but, the institution of philosophy—the promulgation of philosophy—can only be justified on the grounds that it on balance is of more benefit than a waste of time or a deleterious activity.

Human life when it is natural and authentic to its living nature is real, true, good, fitting its purpose and self-justifying. When human life commences to move and proceed away from its authentic nature, it is with that emergence of 'civilization'—alienism—that philosophy gets its emergent, legitimate justification with its emergent purpose which is to struggle against the progressive move of human life away from its authentic nature, being, and reality. Western philosophy has failed to even recognize its legitimate role. And, any justification rests upon its ability to demonstrate that it is, in fact, fulfilling the original, authentic purpose of philosophy. Philosophy's only justification, its only legitimacy and valid purpose is to champion the natural condition of human authenticism in its struggle against its destructive antithesis, alienism, and its general manifestation, 'civilization'. In order to do this, philosophy must experience and receive the human revelation which reveals our natural, authentic, human condition—human authenticism—with its structure-function-dwelling of natural, authentic, human selflifeworld which simultaneously clearly reveals our opposition, alienism, and its destructive, anti-human nature. Paradoxically, philosophy can begin to dissolve away now that authenticism has revealed what human life should be and how it can be regathered: some of us can become progressively less philosophical about the unacceptable puzzle of quasi selflifeworld as we untransmute it back into real, natural, authentic, human selflifeworld.

In short, philosophy's purpose is to defend and preserve the true authentic purposes of human nature, being, spirit and reality; now, as the Human Revelation reveals, which purposes are belonging, fitting and authentic to our human nurturome; rather than experiment and theorize with artificial ideas and phenomena which are revealed as being outside the human being, nature, reality, identity, spirit, and selflifeworld, natural and authentic to us.

## 40. Linguistic Philosophy

Linguistic philosophy throws itself upon the alter of language, not knowing that more is revealed of human being, nature, and reality in the eyes, gestures and movements of mates, family, parent and child, soul mate friends, hunting, gathering, creating tools, utensils and clothing, ceremonies, rituals, rites and dance than is only attempted in language. Language is *a* key feature of humanity not *the* key feature. (*See also* Ch. IX, Part M, "From *Walden* to Nature Human Village.") Language is a functional organ of human reality and being, not the heart or soul of these. Language is an attempt to capture and portray what can be only briefly and partially captured and portrayed. If language was the central essence or distinctive feature of our humanity, where would that leave the above human essences, and others like instrumental music, dance, ceremony, drum beats, sighs, expressive cries, humming, laughter, touch, body language, facial expression, love making, bird song, creek song, wind song, rain song, morning song, night song! Linguistic philosophy is dethroned by these; and, also, by the experience of having heard a beautiful instrumental many times, been moved deeply into our being by it, and then we hear it with the lyrics, sensing that what we have gained with the lyrics which fix and restrict meaning is out weighed by what we have lost of the larger more free inspiration.

Language is one human organ (or instrument) within organismic being, spirit, meaning, communication, purpose, and the sharing of these; one essential tool. It is not these themselves nor the heart of these. Language is the mind using voice sound to manifest a part of human being, spirit, nature, and reality, and to represent a part of these. Both the manifestation and representation are imperfect of these within 'civilization'. But, before the divergence of man from living nature, from natural, authentic, human socioculture, i.e., within natural, authentic, human selflifeworld, language was perfect, albeit brief and partial, manifestation of authentic, human nature, being reality and spirit. For, then, it was natural, authentic manifestation of 'inner' personal elements-phenomena, and referral relating with natural, authentic elements-phenomena of 'outer' human lifeworld: precisely speaking, it was a natural, authentic element-phenomenon within our human structure-function-dwelling, selflifeworld.

One aspect of the post war period can be given much of the credit for the rise of linguistic philosophy and philosophy of language: it is no coincidence that the rise of language study coincides with the information and communications explosions. Scopicized understanding true to its tradition and nature focused upon the largely unstudied particles of this dual phenomenon and another phenomenon that was interjoined within it—the exploding language phenomenon. The communications/information barrage occurred just as language itself was increasing through subdivisions into specialized jargons. This stage in the breakdown of human understanding, communication, consciousness and reality was met by that 300 year old escapism, scopicized inquiry and understanding, i.e., a study of the heart of the constituent particles, e.g., of language, as earlier of knowledge, and matter—rather than a study of the heart of the problem. More widely, when human understanding became troublesome, science studied constituents of matter, mind (psychology) and society (sociology), philosophy studied constituents of knowledge (understanding itself) and the method called rationalism; when communication became troublesome (through barrage, language pollution, language subdivision and multiplication) science (linguistics) and philosophy (linguistic philosophy) studied constituent particles of language, like vowels, consonants, letters, phones, and words. And, when meaning of life became troublesome, ontology looked for constituent particles of (human) being; metaphysics, for particles of reality.

# 41. Gadamer's Hermenuetics

Gadamer's hermenuetics continues to separate man from living nature. He, like all who fall short of universal, timeless human awareness, has his horizons fixed by and dwells within his ideological, perceptual, 'cultural' milieu. The post war focus or obsession with language on the part of philosophy fixed Gadamer's important new view of understanding—new to modern philosophy—within the modern valley of language; just as the focus upon knowledge had for the 300 years previously, fixed it within that valley. The culprit perception is that because only people are constantly talking, listening, reading, writing and broadcasting in the post war, conspiciously linguistic valley, that the language phenomenon itself alone or as leader is what removes and elevates man out of the natural world and defines his new nature—a new nature essentially or generally outside of his older Earth-folk a'civilization nature. Knowledge (or rationality) as the defining characteristic of man, hence was replaced by language, the conspicious and voluminous (within the communications explosion) phenomenon by which he relates knowledge and meaning. Thus, still another illustrious shining feature of man blinds him from seeing fully his own complete nature, being, identity, and reality.

These modern visions we owe largely to one human invention, scopicized understanding. Does not the microscope intensify light upon a given area leaving all others shadowed in darkness? Gadamer having escaped for a revelatory hour, from the spell and error of modern science, to behold that there is an elemental understanding and awareness that comes before science, upon which science rests, and to which science should function as an augmentation of; then, ironically and mistakenly he returns to both constituental understanding of modern science and its milieu of linguistics to assert that every mode of behavior to the extent that it is meaningful has a linguistic foundation; "being that can be understood is language." Such is the powerful pull of ones own age and milieu that a true revelation can be followed by a stumbling return into huge shadows of conventionality.

## 42. Relativism

The view, central to contemporary relativism, that all observation is 'theory laden' can now be readily dismissed. This amounts to holding that prehistoric and historic man's observations have all been attached to a theory rather than to the understanding through sense, instinct, reason and intuition of intelligence and its natural law. On the contrary, men always have known intuitively or by whole, inherent intelligence that falling fruit, falling tossed spears, and himself stumbling and falling were caused or guided by an intrinsic force within living nature. They knew and experienced this force as pure natural law, not theory. Theory arises with science while observation and understanding predate both. The relativist should rather say 'all *scientific* observation is theory laden'; which one hastens to add coexists even today with *prescientific, natural* observation and understanding. To state that law is merely correct theory is to have it backwards: correct theory becomes, attains to a fitting with phenomenal, natural law.

Which brings us to the modern view of science that the physical sciences attain to laws, but that the social sciences can only attain to theories, however hard and reliable a few may turn out to be held. Here the revelatory nature of authenticism penetrates science with this light: since man is rediscovered to be not just a social animal but an animal consisting of a natural socioculture and nurtureculture inherent and authentic to him, social theory can now attain to natural law, i.e., the laws of man's natural authentic socioculture–nurtureculture and selflifeworld.

Relativism in moral and social philosophy, and philosophy of science and social science holds that thought is true and human behavior good only when fitting within a harmony of (in relation to) a particular socioculture, whether simple or complex. This relativism is a partial truth, and, therefore, generally a falsehood. The whole truth accords with an objective human reality: the revelation that our (c.50,000 year old) human subspecies belongs to/with a general kind of socioculture distinct to pre'civilization', and a'civilization' Earth-folk socioculture, and distinct and apart from 'civilization' socioculture; and consisting of the original, archetypal elements-phenomena that are authentic to our species and that are universal and general in a'civilization' Earth-folk socioculture. The exceptions to this are two kinds of non-general, experimental elements-phenomena: those pulled from living nature for experiment, and those penetrating in from alien realm—from alienism/'civilization'.

# 43. Mind-body Unity

Contrary to popular view, the mind is not the sole source of consciousness but is merely the center of it. The essentials of consciousness cannot occur without the senses and their bodily organs; nor can essential phenomena of consciousness, e.g., walking, grasping and perception, come into consciousness without the body parts, e.g., legs, hands and sense organs required for these phenomena. Consciousness, then, is organismic; and, it follows that the mind also has these sensory and phenomenal essentials and is likewise organismic. The mind is not, therefore, the brain which is an organ. Both the mind and the brain are central processing terminals: the mind for organismic consciousness, the brain for the organism. The brain, then, is organic to the organism; the mind is organismic *with* consciousness and consciousness with mind.

There is no such thing as the mind-body of traditional philosophy; there is only mind-brain-body because the mind-brain cannot exist—except towards sudden death—without essential parts of the body, e.g., glands which supply continuous essentials, e.g., chemicals and enzymes. And, since the mind and consciousness cannot exist without the essential, constitutional phenomena of human being and spirit, there is mind/brain/body/spirit/being. All else is scientific reductionism or philosophical reductionism.

# CHAPTER XI

# THINK VERSE, SOUL LINES, ECOPOETRY

**Ecolove: Ode to True Humanity**

Why do we love thee, old Humanity?
That we may find the ways
To clear our fields of being.
We love most deeply
The you that might have been spared,
Searching for the us
That pure Living Nature meant to be.

But, then, we love
What you are, and will be,
The struggling you within us,
As we push through
Our Age struck by a storm in time,
The pollutions of our culture,
The miscarriages by history,
Their misfortunes for personality
That suppress the growth and fulfillment
Of our hearts and souls.

We love thee
As we both have left
To go the real journey,
To hunt and gather the sweets
Buried by the debris of exploding novelty,
To devour these lost nutrients,
To reclaim some growth-destiny denied.

We love that we fight the last fight
To re-grow whole self and being,
To cleanse and heal the Home of Life,
As we soul mates
Struggle to purify senses, mind, and living,
As our spirits recover lost union
With Earth's Being-grace.

## Alien From Nowhere

How long this beer can has been here
Is what the human mind first ponders.

Wedged in sand and rocks of washbed
Fifty miles from its world of nowhere.
The store that sent it loose in human hands
Occupies a lot, in nowhere, to these shifting sands.

Judging from its faded luster and bent condition,
Its been here maybe twenty years, not twenty million,
Alien to this bed carved in mineral and life,
It has no role in this play of wild fruition;

It speaks of 'civil' beer where none will hear.
It hurls questions piercing hope for reconnection
With one resurging, primal pilgrim pausing here.

Destined to be silenced, but for a moment
It names the creature that dropped it here;
It answers the last question it raises:
"What creature breaks this joy with crumpled spear?"

One of no participation: one *Homo alienus,*
Alias Kilroy; but "*Killjoy* was here."

## Quakes

We know well of quakes upon the Earth.
But, these sudden moves within our soul?
Violent moves beneath the crust
Of our accumulated deeds called culture?

Does God or other will divine
Relieve this pressure upon the soul?
Defend this child of pure creation?
Redirect our steps in sands of time?

Does force of life, *elan vital,*
Wipe our deeds of arrogance and greed
Off some map of primal life and soul?
To heal and fuse the malnurtured and abused?

Does all life and being *as one* preside,
Singing a deep spirit song
For all beings' ride through time?
Does it shake off convention and pretension
For recollection of that song forgotten?

## To Speak the Body of We

Language says to life,
Give me your sounds of atrophy and pain,
Your sounds of growth and joy,
Your sounds of struggle to be—
Free of pretense and greed—
Sounds of your primal, pure identity.

Give me sounds of life attaining being,
Echoing round and on to speech
As life evolves its charted strife
With soul, into self-conscious being,
As caring leaves the self of I
Joining the cause of We, in social life,
Searching to speak our needings' Why.

And the body of life and being,
Organism-Earth, says to language,
You are not needed here,
But are welcome, if you come to speak
More deeply than, "Listen to me":
To know the condition of your being,
*To speak the body of We.*

## Junk World, Flash World

See the Earth becoming a junk yard,
Disguised, supervised by a throw away mentality,
Clinging to ego-life's desires,
Throwing out beans of Life
With the shells of a culture
Built only for its flash in time and being.

See us flow in history's shallow being,
In the river of life leaving its banks,
Flooding us beyond spirit-soil to empire hills—
To imperial power girded with steel and electrons,
Where sterile housing tracts and media towers
Scarcely tap Earth's spirit-juices,
And flash out the electric message—
That electronified mind can electrocute
The Old Spirit, and create a new 'Second Nature'.

See babes torn from arms and teats,
From wild things, touchings, growth sucklings.
Yet, Dick and Jane wade a creek, climb a tree,
Seeking, clutching for the lost Tree of Life.
See adults stuck with sci-technic toys, commercial pampers.
While deep thinkers are searching the wilderness.

See pipes suck water through the empire hills
For a culture bleeding nurture
From geometric seams and mathemetized schemes
That deny deep Life, soul, and meaning—
And deep destiny from a gorge called Olduvai,
Through a long, green valley called Eden.

## Fossil Speech

Sounds blown from soft, red flesh,
Through the box of voice sound,
Through the cave of mouth to taste
Sounds that search for form of speech.

Words so stubborn, so embedded,
As to be petrified by time
Into chanting ridges and calling buttes,
Defying storms of change.

Speech sounds given as kisses defying death,
As mouth to mouth spirit resuscitation.
Deep language carved into stone
And etched onto sheaths of wood,
Buried as scrolls in caves, as bios in tombs,
Carried on culture's beasts and vessels,
As treasured pages of Ages.

Speech as human spirit's reach
To endure every breach by culture
Of nurture passed as birthright to all and each,
As soul-life sounds midst bones and fallen stone.

Sounds regathered not for science shelves,
But as a forage for a Soul
To speak again what is older,
What is stronger than constructed stone:
Speech calling from the fossil Soul.

## Story of the Night

Modern minds know of outer space.
The telescope, radioscope, and such,
Have given us the cold scoop!—
The sky we saw for ages,
When telling stories as firelight died,
Was not the real sky
We, now, are privileged to know!

Yet, Human eyes have seen
The sky for a million years!
Eyes of ape for fifty million!
Which is the real night?
Do artificial scopes give the story
Of this, Nature's own creation's?
Does life evolve such mute deception?
Such refutation of *creation*
Made and made known
*Through Her perception*!

Wolf still howls his story of the night.
Seems to mock our lost wildness at such sight.
Are we the worthy measure of all things,
If we bow to edicts of machines?
Does reality rescind as perception bends
From penetrations by machines?
Or, does it stand the test of human time!
And past a test of human being!

What is the *human scope* of things?
And in *that* lost scope

What was the story of night?
Are flying, star-stuff theories true?
Can distant data add up to
The awe and bliss that fed our human story?
Do the stars confirming configurations
Reveal the ancient, human spirit's move
Through the time and space of Earth!

If eyes are windows of the soul,
Through which it knows the world,
They alone once saw the night,
And flashed that light to the soul
For *its story of the night.*

## Leaving Her

Once we were not alone,
Our flashing, global tribe
Of electric spears and speech.

Now egoism conspires to wield stolen power,
Divorced away from our Earth kinships,
From the missing sojourners of Life
That took our same paths of seasons,
Followed Her 'reasons'–Earth places called *Home*.

Once, we lived together
In community of Life–
The journey of Being through time.
We made offerings, deals, kept treaties.
But it came to pass, we parted.
We left Her communal journey through time.

We parted from Earth-life,
Taking life through an artful time and space,
Leaving our source place,
And its road long-traveled by.

Making no vows, we keep no union with Her.
We live personal want and desire.
We rule all land and beings.
The deal is made to kill our primal needs
With a technics' 'Second Nature'.

Life still calls to us, beseeching us.
But, we hear the others no more.

We speak no tongues of Earth-spirit
We learned Greek, hail Plato!
We learned Latin, hail Caesar!
And Indo-European and German, hail the British Empire!
And Commercialean tongue, hail Egoism's imperium!

We move no more with Earth-spirit.
Damn Her and that old dance!
We take Her to our party of culture,
Show Her dances she can't do!

We get Her soiled and drunk, and to bed.
And all these bastard offspring issues
Make the enterprises for intellectual adventure,
Make sport as Will gets drunk with Ego.
Building empires for Egoism's greed.

See Her housebound! Confined to bed chambers!
Soiled tramp overrun by bastard kids!
Our hole of desire
For Being's fall through time and space–
Our fall from Earth's Being-grace.

## The Question of Being Alone

People awake with sighs or moans
And see the artificial rooms we've made,
And hear the sounds of electric technics,
And ponder the question of lone being, of being alone.

Some scientists say we are alone.
Some, searching the skies, say we are not.
But, the rooster and a thousand other critters
Announce to all sentient souls of Earth,
That another day is born from night.

But we do not see the Others awake,
At break of day, a hundred social mammals
Announce the light in their being's way
To all old souls–ah, even we–
That precede man-made life in rooms

Sculptured walls scalpel our dissection
From those Beings with once we grew.
Yet, pure soul dreams awake us to see
Through the modern myth, *We are alone,*

Through the cyborg-Cyclopsean searchers,
Through their mechanical eye blind to spirit-world,
Through their hope to announce, *We are not alone!*
And through the illusion in the announcement.

## Pretty Lie

They say a picture's worth a thousand words.
If so, then why do poets write?

Do techno shots on paper and on movie screen
Lack what science and art imply they mean?

Is this magic show a seduction without end
To hide some lies on which we now depend?

Does the poet's eye of spirit-mind
Refute the eye of science-gadget mind?

Does this condensation of our speech
Reveal some truth that science cannot reach?

Will the poet rise up to defy
The rampage-mind's illusion, that *Picture's don't lie.* ?

Will the spirit-eye go blind
From techno-eye, so prettily unkind?

## The Bang Within

Culture breaks its boundaries,
Language spills over its banks,
Flooding the fields of human being—
Plasma hemorrhagings of culture,
Spilling spirit-blood from our mating fury,
From child's abuse and nurture's crush by commerce,
From torn family tendons, community organs
and Earth places,
From the human face struck by flying space,
As the sky is really falling,
As genes and culture are torn apart
By the explosion from within,
After human stumblings are compressed and suppressed
By strokes of ego-genius thrashing out,
Thrusting into nature's time and being,
Igniting an unseen, last, critical mass,
Hidden in the loins of our soul.

What are we now flying outward—
Through our self-created, *artificial, wilderness* 'frontier'?
Suppressing a last primal scream
Stuck unheard in throat's phlegm,
Untold in mouth's foaming, language madness,
Unseen in vacant eyes program hypnotized!

A scream too deep for body's orgasm to release,
Too deep for mind's gushing formulae to reach,
To break the fever of our soul,
Deprived of Living Nature's mind,
Severed from will of Magna Mater!

Now flying outward, our egos embrace intellect!
We study debris for a reasoned reassembly of self-life-world.
Dripping and drenched in Earth-life blood!
Bleeding juices of Life's creation!

**War and Peace**

To those possessing nature's patience, death will come:

Who wait out chaos of two storms,
One roused by nature, one from artificial norms;
Who keep the steady rounds of nature;
Who know her unspoken nomenclature,
And language deeper than the civil tongue;

Who know life belongs to her selections,
With no false frontier into which to race;
Who have faith in time and being's predilections
In realms predating the scientific pace.

Who know self as not a thing of one,
But a graceful union of all that will forthcome;
Who know life's ingrediential fit,
In the end, will outlive our modern wit.

To those deprived of nature's knowledge, death will run.

## The Right Stuff

What to do with all this star-stuff?
Coming at us on home screens from scopes!
Go outside and try to find some!
In the yard, down the road, on the mountain!
Only images on a screen from a scope!

Outside the scope, the room, the city,
We find Earth-life in retreat,
Eke out our quasi lives
Within the dying Spirit of life,
And fashion makeshift family and community
From a blend of the artificial and the real.

Star dust, exploded by sci-technics, blasts imagination!
Mind is buried alive, doing ego titillations,
Trapped outside the body-spirit stride
Of our Human Being's ride
On the thrust and pause of time and culture—
The spirit-purpose world of Earth-life!

Science blasts star stuff down upon the soul,
A fallout cloud upon our schools,
A toxin inhaled by family-community.
Spins off technics feed the growing rich!
No new-found lands for the growing poor!

But, Earthy spirits defending our home
Would launch star-stuff back into space,
And fire a human message
To the mechanical-eyed scientists:
We were created from Earth-stuff!

We were born with right perception.
Endowed by creation with the right senses.
Our eyes have always seen the glory!
We were made of the right stuff!
Earth-stuff exploded with Life and us!
Our station of Life *is here*!

**Recent Sex**

Two new sexes have emerged,
Just recently by primal time,
Their mating rituals couple violent passions
In desperate union widely known .

A marriage untested, still young—
Written words in union with idea.
While pure idea forswears the union.
And their children's fate is still unknown.

Just why speech eloped with written word;
How idea lost its copy of original reality;
Such went unrecorded by history's tongue.
How long we will now speak, cannot be shown.

## Story of Earth-Life

Before Earth-life, there was only sky,
Sky of star-stuff flying outward with no end.
Then, planets formed around the close star—our Sun.
And the Earth, receiving the Sun's light,
Became with Life, and bore it forth.

Life made a home to play in, live in and share.
Many Life forms were born in Earth's world.
They found homes: in colorful communities
Of plants and animals that needed each other.

And Life became with feeling,
Bearing forth sentient mammals—
Intelligent and social–like dolphins and chimpanzees.
And the deepest feeling was the need for love.
And home was the place where Love began.

And Earth-life needed time, to live and love.
So time emerged with life in sacred matrimony.
Life, time and Earth lived and played together almost forever.
Then one playing animal, playing through time,
Played more and more like children play today.
And the wild, free play of Life discovered people,
And they were given a part in the beautiful play.

So Life was on its Journey through Earth's time and Being.
And Life with feeling loved its Earth-life-world,
Its sunlight by day and starlight by night.
No dead star-stuff! Just wonder and awe of firmament,
And work and play with Earth's Life-stuff.

Life danced, and moved with, and spoke of, Earth's Play of Life:
In Life's goodbye march away from dead star-stuff.

And the Journey of Life moved on through eons of time.
And Earth came to be Life's home for eternity,
Bursting with millions of plants, animals, and then people;
Each kind dancing and singing in the play created for all,
To live, play, and love until the end of time,
Until the end of Earth-life and its Journey.

Earth was paradise Home for Life as old as the hills;
But something happened to Earth-life and time,
Making time run faster and faster–towards its end.
And the people saw time ending for many plants and animals.
Some worried that even the people's time was running out.

Some people missed the dead dancers and the missing songs;
The play was less beautiful and good, with players dying out.
Others did not care about the loss of dancers, of such beauty lost.
They danced mostly with machines since progress deems *That's life.*

Some left Earth-life behind to live on dead star-stuff!
Stealing Earth's Life-stuff off to a dead planet!
They recreated themselves with computers–
Emerged aliens to Earth-life-world!
Accepted the end of time for many plants, animals and Earth people.

And it came to pass,
As the people came to be more and more alone,
They began to ask, "Are we alone?"

## Retaken

How long this trace of road has wandered,
Dying, now, from lack of human reasons,
From sieges of growth, burrowing, and seasons,
Is what the human mind first ponders.

And what became of the dream abandoned
That rebelled, and took this path to somewhere,
Hoping for no trace of someone having been,
For chance at grasping *deeper self* now here remanded?

And how long this dreamer's dream,
The resprouting of a life and soul?
How much of being's lost ground retaken?
Is what the heart would the mind could know.

**Perception's Storm**

There is not much coming in
Through the poet's adulterated mind,
Through the brave new world's scientific fabrication,
Through the storm-tossed sea of information,
Through the gloss of image and sensation,
Through the view of novelty as creation.

There is not much coming in,
As this infidel, this hurricane of time,
Batters senses, once kept pure, for perception,
As artists surf waves of brash imagination,
As their 'second nature' takes formation,
As souls fight tides of alienation.

What sea reclaims this species' nations?

## Machine

We believe in the power of thought,
That intelligence determined what destiny wrought.
But the machine is now king philosophy;
Its verdict on nature, unquestioned as true!
Its sentence of being, deemed right and free!
Technique is our fulfilling prophecy!

The past is but the history of mechanical zeal.
As reason surpasses blood-truth we feel.
As will and purpose are whatever is new.

God is but science behind modern actions.
As world is realm of our mind's refractions.
As life is the oiling of incredulous sqeaks.
Being, merely mind passing through contraptions.
Time, but the history of Ego's techniques.
Nature, the boundary around which it sneaks.

## Flow

Ponder the phenomenon of *flow*:

The flow of blood to the cells,
To the brain, and to the body it propels;
The flow of life through species' season,
As time and being debunk all reason;

The flow of breath through the flute,
Our answer to the bird, we are not mute;
The flow of fingers upon taut strings,
Our call to bird, our soul has wings;

The flow of hands caressing flesh,
Fulfilling touch, soul's first tenderness;
The rhythmic flow of bodies mating,
Timed release of life-force waiting;

Forget the first blast of fire and gas,
Over-lauded birth of indifferent universe;

Forget all volcano blasts, of long past;
Remember lava's flow into molds,
Magnanimous mediums where diverse life holds;

Think of flowing clouds and air,
Storm water the Earth could share
With land and sea, and life that learned to care;

Think of speech as a flowing brook,
A stream of vocal consciousness,

Once, in time, still running free
Of laws and lines enforced by any book;

Think each species a rivulet flow,
With rushing gush of instinct to know
Of contributory being upstream and below:
See all beings in a watershed whole.

Know the flow of nurture
Was once at one with culture,
Till blocked and channeled by desire,
With time and being pooled out for hire;

Forget science's arrogant apparatus-vision;
Remember self-conscious being
Flowed from nature's own divisions;

Know the sacred is on Earth, a time and place
Where life achieved the flow of grace.

## Old Knowledge

Leaving the house
To fetch sun-dried clothes,

The soul pierced by bird's call
Knows the wild is on call;

Knows bird sings to all life
That knows a song of being;

Knows mind caught in a storm of time
Can be calmed by ancient will;

Knows the wild within survives
To slip back into eternity;

Knows being can be recalled
Into its current of time.

**New Knowledge**

Too much knowledge got in the way
Of what the soul would starkly say;

Culture forced perception thru technolgic storm,
Eyes, ears, and touch took artificial form,

Speech, humanity's mark, is electronified.
And *science as final God* has been codified.

With experience vicarious, and relationships programmable,
Reality becoming virtual is the ultimate gamble.

Hitched to reality, life will be virtual!
Can the soul hide from this scientific crucible?

## Through This Skin

The ultimate expression of art
Lies in the eruption of the beauty within,

Trapped under dinosaur plates of convention,
Culture's counter-evolved counterpart

To Nature's primal, human skin.

## History

History? Five thousand years old,

Scrawled symbols so elegantly bold,
To revise and bury the old,
To suppress a primordial theme
With constructions of greedy scheme.

The 'civilized' brute is mean!

## Being's Memory

Some won't ever forget
How our Soul takes in
Deep breaths of time!

Being somehow remembers—
If only in dreams of sleep,
Or only in flash of revelation–
The deep breaths of childhood time.

The deep spell we fell into,
When gathering wild berries,
When catching fish, or storing food,
Or the deep panting of hide and seek time.

Being somehow remembers
The glowing of spirit, its shine,
From intimate concourse and consorting
With flashes of time.

Being somehow remembers spirit-time,
Before thing-time and machine-time,
When all systems were go
For Earth's space we now know:
Is nest of human space and time!

Being somehow remembers a time
Before Ego's thrust for power
Released the great flood
Of automated bodies over concrete,
Before urban, commercial feat

Crowded out the rituals of mealtime,
Part of our Spirit's move through time.

Being somehow remembers a time
Before speech suffered culture's breach
Of pure words from a primal spring
Still flowing in the loyal banks of time.

Being somehow remembers spiritual intercourse,
When Soul, sprouted as family saplings,
Caught an ancient breeze-song
Whispering, *It takes a village.*—
An ancient creed now gone
With the stormy wind of a flashy time.

## The Answer

Does the primal Human Journey end
At sci-technic's stay of execution?
Can we carve our Soul some retribution?
Will the modern dynamo ever end?

Was human nature once divine,
A character lost in "second nature's" design?
Were authentic lines we used to say
Torn out with the mutilation of the play?

Is our lesson beyond the wall and fence?
Out in the wild with all lifekind?
Can embrace of being send arrogance fleeing,
Recasting our part in the play divine?

Why does science embrace growing, endless questions?
Yet, scorn the questions of *Why?* and *So What?*
Why does art celebrate its endless creations?
Yet miss primal rhythms of our Soul's regeneration?
Is the answer flying through millions of years—
Feeling, conscious Life? Its joys and tears?

Are we Life's human flowering?
Picking other flowers for nonchalant devouring!
Are we, electron-steel-human weeds,
Choking out nature's finest deeds?

Will human works leave little trace
Of Earth's answer to exploding space?
Do we so ceremoniously doubt

The answer Earth so patiently awakened?
Will we ever shout the answer out?
Will we ever call it in
To the human Soul foresaken?

## Unaborted Lines

In fresh penned lines of prose
May lie deposited a poem.
One can feel the kicking
Within the lines that send
A refusal of deliberated end.
A playing hooky from the school
Where formal mind doth rule.

Such mind's not life itself;
And, one law of life doth say
That if not abandoned,
This soul's rebirthing labor
Carried to terms of the heart
By a different drummer,
Will sound a rhythm to savour
With the birth of poetry's art.

## Flash

We know when immortal lines
Escape constructions of the mind.

We know even common man
Can feed the fire of soul,
Send up the leap of flame,
That sends out the flash of light
Into the chilling, mortal void
The arrogant tongue calls culture;

We know when words older than the civil mind
Escape those bounds to flash their light.

## Calls from the Wild Within

Listen! Do spirits lurk
In background sounds?
Do elements and participants
In life's matrix-elegance
Speak the tales of Her Journey?

Do things of Life sing the Earth-life song—
In sounds of wind, trees, birds, beasts,
In children's play, a learning feast,
And people working tools and soil?

Is this a choral story–
Told in nurture's waves:
Of acts of crafting clothes and utensils,
Of preparing food and suckling babes,
Of gathering food from wild, pure dells?

Is history Ego's dams on the Soul's river?
Do lakes of science damn us—
Drowning much of nature's sounds,
With machine sound, and airwave voices,
And deny The Way of Spirit-things!

Is there a song of Our Self?
Does whole Being have needs?
For natural, background sounds?
And surrounding things of Life?
Is Our Soul lacking such fulfillment?

Do natural Souls hear a primal song?
Do they have a natural, dwelling place?
Are we just migrants—leaving home?
Does Our Soul have community critters?
Are we in mutiny at that bounty!

Is the Wild calling out to us?
Is that perfect poem or song, clutched by memory,
calling back to song 10,000 years ago?
Is Soul calling back and searching–
Gone fishing in the spirit-gene pool?

## Oh, Fresh Eternal Day

Oh, flooding day, you measure our soul!
Refreshened by night always near,
Each soul may chance its reawakening
With steps into your overflowing sphere.

The year's nothing without the day,
While the day's so much unto itself,
As moment on moment the soul's fulfilled,
Or stalks escapism for fleeting help.

Years note being's encircling seasons;
Days channel our heart's flow of feeling;
Clocks mark hours as triumph of reason;
Calendars mark our eclipse of being.

Life each day rises to your light,
Bathed and cleansed by dreams of night.
We face light or cower in pop culture's shade;
And how is this light, now, subject for debate?

## Body Meets Culture

New baby shouts out, head first,
Through wide stretched body lips,
Out of the mouth of womanhood.

Before first breath screams *life!*
Young girl shouts at youth's oppression,
"Now I am woman, I am somebody;
My baby's cry is my cry.
Trapped between girl and woman,
I break my culture's chains,
I was I, now I am we.
I was lost atomistic ego, now I am whole and free,
Having seized my inalienable rights
Denied by culture's newborn world, adolescence,
My body shouts the proof: it is phony!"

The body has spoken, striking a blow,
And culture replies, striking back.
And this is to each again and again,
A dialectic, by appearance, without end.

Culture counters nature's equality
With unequal nurture and environmental opportunity.
The mind and body are split
Between temporal expedience and primordial fit.

Unwed mothers so desperate to be,
Are your babies, then, born free?
Free as wind blows, as grass grows,
As wild horse and wolves run

To receive eternal joys and pains
Presented pure by Magna Mater.

City has spoken the cultural myth
Out of mathematics' precision,
The sci-technic dream to explore
The fragmented world we inherit
From the illegitimate phallic thrust
Splitting life away from spirit.

Mind has shouted to living soul:
To babies hooked to meds and machines,
To children hooked on wired and wireless screens,
To the unwed mother's fleeting, victory dreams,
"Forget Living Nature's primal spirit!
You shall not know Her fields of beings;
*You shall obey me!*"

## Spark or Fire

Are we a spark–
As outward minds aspire—
Searching a cosmos vast and dark?

Or is our Soul a fire,
A passion of Being raging with Time?
A rhythm outside of scientific rhyme?

Are we embers of the stars,
If after volcanos are cold,
Life rose from such rainswept molds?

If all passionate affair
Burns from Earth and Sun consorting,
If all joy is from sunlit waters near!

If from this homely birth,
Our soul emerged from that strife,
Can we embrace this fire of life!

If this fire against eternal cold
Is tended only here upon the Earth,
Can we live this story of our Soul?

### Life and Soul

When life evolves to that place
Where self and life interface,
With Soul and world perceived as one,
Time and being become as one.

The sacred matrimony once in place
Holds to its patterns and its pace.
Novel things break in to have their way,
Forsake such bliss for parties of the day;

Wild oats of Ego fly at the face
Of downtrodden youth still yearning for its grace.
Greed and lust run naked in the street,
Till life confronts the Soul it has to meet.

Nature waits out the wars against her will;
The springs of life from tears again refill.

## Forgiving Earth

Earth, mother of Life
Has been forgiving of anti-life.
Repeatedly she has nursed
Oppressed and stunted souls
Trapped stratified in city's mounds,
Human mortar for power's art and stones
Set free repeatedly to be whole again.

But, what do egoistic mortals know
Of divinity in Her life's eternity?
Of what might have come to be:
Free of human illusion and delusion;
Free of kings posing as gods,
And families royal thru fabricated supremacy;
Free of religions worshiping
Their singular success within historicity.

Who knows what might have been
Sans the mechanical thrust of 'civilization'?
What love has been lost in subjugation,
What songs unsung in regimes of regimentation?
Who knows what joy, what flowerings of soul
On Her human fields of being
Were drowned by dammed up lakes of greed?

Will Earth, mother of caring being,
Remain ceaselessly forgiving
Of anti-Life and anti-Conscious Being?
Will She have a place

For beings that blow away
The intimate relations of the Soul
With probing blasts from copulations
With Her, mother of paradigm creations?
Such acts outside the sacred union they deserted!

Is outer space, which Life rejected,
Reserved for those who alter genes and DNA,
That think power is the grandest, human play?
Can She absorb incestuous violation
And love this child called *human being,*
Who rapes her places of creation?

## Time Affair

Say what we may of time:

How slowly it moves in youth,
Before elements of age quicken it;
    How time kills the sweet bird of youth
    For the sake of the nesting place;
How it settles the fires of youth,
Leaving warm love of maturity's grace;
    How time heals all wounds,
    Yet, scars the spirit–the real marks of aging;
How it mocks our desire to have and hold it,
True to its first Love, passionate Being;
    How time rejects poetic suitors from history,
    Embracing wild, pure Life, in immortality;

How time fells civilizations
For the sake of Life's natural order;
    How science gloats over time and space,
    Yet, only time has meaning when caressed by being;
How clocks, calendars, and mathemetized schemes
Obstruct Earth-life's flow of time and being;
    How for everything there was a time and place,
    Till novel things brought the chaos we now face;
How the more things of culture change,
The more human depths remain the same;

How change rides time, like rain rides wind,
    Even as the winds of change turn to storm;
    How time and being are a sacred matrimony,
Set against our violations by culture,
And the human rapings by history;

How mortals rediscover time and nature don't wait—
For us to flee a cultural storm that binds;
How we, having been with sultry, stormy time,
Must turn to face what might and should have been;
How we find this was a swifter affair,
With a mate less faithful, than we thought;
How we sense a more fair and worthly time
Could have more faithfully borne
Our Being's natural order.

## Incubation Words

They flock with birds of their feather,
With poems as their eggs,
Which are laid through the mouth,
Sat upon by the mind,
Kept warm by nestling hands.

And, if the creature laying them
Has kept to laws of human tongue,
The shell breaks, the poem awakes,
To find its way in language.

## Informed Speech

Oh, essence of poetry!

The knowing of sounds of speech,
Releasing the beauty of voice,
In marked time with sounds of silence

That extend its reach past prose.
To silence the explosive noise
Of history at war with nature!

The breaking free of artificial form
Toward a purity of language!

## Time and Space

Where is our place in space?
There is not enough space–
Geographical, biological, spiritual-ontological, space!
Give us more space, real space!
Inner space, home space, soul space!
Enough for spirit breathing, soul touchings, soul moves!
Enough room, clear the way, give us room!
These weeds! So thick and high!
Can't find the tree of life, nor sky!
Plastic and steel weeds! And the noise!
The noise when shot through with piped fire!
Overrun by the stuff! This whatever stuff!
Where are we? Where is this place?
What time is it? What's the blasted time?
These damn explosions!
Exploding information, communication, transportation,
culture itself!
How late is it? Have they blasted our time away?
Outward into outer space and cyber space!
Has our human time run out?

Who are we! What are we!
Time and being cannot be separated–
Beings flying outward–from time flying outward?
Give the human race a break!
At least lets find us a name!
Humans don't do this–live this life!
Unidentified flying beings–UFBs
Yes! With our bodies stuck in concrete and steel!
But, they are working on that;
A few will live on Mars.
The rest will watch it on TV,
And fancy their children will make the last leap,
Fancy we'll get our bodies reunited with our UFBs!

## Real Exploration

Talk of explorations!
Space, microscopic, technologic explorations!
Yet, has the self shrunk to an atomistic ego?
Has the mind become a Cyclops monster,
When the eye has conquered the brain?

Talk of mystery of origins!
Yet, does the natural Soul just know?
Outside the explosion of knowledge,
Where its been, and where to be?

Talk of mystery of human destiny!
Yet, is Soul retreating from flying debris?
Is living *to know* or *to be*?
When spirit breaks through, to freedom,
Is a storm of knowledge any part of being?

Talk of a new age of discovery!
Yet, must not spirit be the seeker?
Uncovering *what we are* and *why we are*!
Finding knowledge is memory of moments of being!

Is knowledge a grasping of cold facts?
Or memory caressing the Soul's acts!
Is life's journey with time and being
A story of knowledge? Or of caring and feeling!

Where is real exploration? Of evolved, Human Soul?
Where is whole perception as the ground for seeing?
Sprouting through millions of years,
Exacting selections of joys and tears,
Life *feels its way* to fields of being.

## Second Nature, Second Death

Will the mind kill the soul?
This very mind of us
That rides with lust the mind of old,
Old dreams resisting hopelessness?

Mounting primal forms that hold
The union of the mind and soul,
Programmed minds, blinded by desire,
Put seduction up for hire.

The finest minds we trust
Boast of techno phallus's thrusts
Into virgin realms of being.
Can the soul abide such bleeding?

Will this 'second nature' disconnect,
Dazzled blind to its defect–
Doctrine of mind as ground for being–
When the soul lets go of make believing?

## Body Time

The body knows the time:
Has its organismic rhythm, its ecologic rhyme;
Informs the mind when to sleep,
Awakes it to the body's time for motion,
And when to rest its locomotion.
To heart and soul, it has promises to keep.

The body makes its rounds.
Gatekeeper of the soul, it has
Well worn rhythmic, circling paths:
The trusted part of being's grounds.

The modern mind desires no bounds.
The modern pace shows on body's face,
And on body's wrist and favorite walls
Clocks appear, inventions designed to erase
What the body's promises hold dear.

The modern mind would force its will,
Press its doctrine the body comes to fear.
But, the body slaps the mind with pain
To remind it what is loss and what is gain.

The proof the body is not dull?
It trapped the mind within the skull!
Forget commercial, futuristic slogans!
The mind lives off the pull
From the senses, the web of life, and the organs!

## Old Soul

We decline to believe in social guilt,
Sparing conscience, and society our ancestors built.

Yet, the belief holds out, we are born in sin,
Fallen from grace, but can be forgiven,
To enter into a next world and life;
A religious escape from our evil and strife.

Forgiveness, then, must be the key
To the shifting of guilt from I to We.

We get a forgiveness from a God quite young,
Born when the written word began to be sung.
Our belief in hereafter is so dependable
Our souls on Earth are quite expendable.

Oh, where's the faith in old Soul of Earth
To forgive, to redeem us; a second birth?

## That Thousand Years

In the mightiest human places,
Celebration on billions of faces;

The cleverest and richest nation
Wait the collision of prophecy and calendar time
With their latest, transforming creation.

And when that thousand years passes,
Prophecy is down, and the machine still rules.

Dodging their own bullet, these fools
Still prove the most clever of asses.

## Starkly

The poem lies so naked on the page,
Baring memory, insight, love, or rage.

With no apparel of prose to warm our way,
We touch what a soul came starkly to say.

## Resolution

New Year's resolutions are anciently real:

Knowing the soul is growing, constantly verging
On predictable, new growth emerging;

Into this surge, toss the conscious will.

## Youth Triadics

When we are very young,
We know the key to life
Is a quick mind and tongue—
The way that love is won.

A little later when in school,
We realize another rule,
The moving body is our friend—
In structured play, for hearts to win.

But, then, in adolescent days,
The quick mind, body, and tongue
Meet the new, triadic unity of fun—
Of quick body, *lips,* and tongue.

## In Memoriam

Let us start the day in contemplation,
That waking door to nature-human relation.

Reading will be disallowed, a recent thing,
First taught in the circles of a king!
Why break the primal chain of being,
Its loyal links–contemplation, conversation, and relation?

Reading is the first, bad magic,
The first curse on the will,
The ride on outside thoughts that preset
The first addiction to a pill—
The addiction to word capsules of an alphabet,
Symbols grouped and strung together, so arranged:
To assure the mind will be deranged.

Nature blesses those that cannot read,
And those who quit this addictive need.
For nature's thought, feelings and speech,
And soul in synch with mind before pen's breach!

We are pushed very young into addiction;
Our graduation: 'civilization's' benediction.

## The Year Recalls in Thee

These yearly turnings, like October to November,
What would matter of such periodic ends
Without the turnings of pen on paper
To reach those the year portrays as friends?

What a year of human being most recalls
Are gestures from the heart toward its eternity,
That part of it that harbors soul
From the storm we call modernity.

Ask not if Rousseau's 'noble savage' really lived.
What reaches from instincts of the heart,
What defends our species, our soul's eternity:
This nobel 'savage' lives in thee.

# NOTES

Chapter IX

1. Neal Evernden, *The Natural Alien; Humankind and Environment* (Toronto: University of Toronto Press, 1985), 22; sited in George Sessions ed., Deep Ecology for the 21st Century (Boston: Shambhala, 1995).

2. Max Oelschleager, *The Idea of Wilderness* (New Haven and London: Yale University Press, 1991). Emphasis added.

3. Paraphrased from an unlocated source. *See* also Delores LaChapelle, *Earth Wisdom*, 1984, and *Sacred Land, Sacred Sex: Rapture of the Deep; Concerning Deep Ecology and Celebrating Life* (Silverton, Colo: Finn Hill Arts, 1988).

4. Paul Shepard, "A Post-Historic Primitivism," in Max Oelschlaeger ed., *The Wilderness Condition: Essays on Environment and Civilization* (San Francisco: Sierra Club Books, 1992).

5. Ibid

6. "Walking," 224.

7. Michael Zimmerman, "The Blessing of Otherness" in Max Oelschlaeger's *The Wilderness Condition: Essays on Environment and Civilization* (San Francisco: Sierra Club Books, 1992), 264, wherein Zimmerman seems to largely agree with "Hans Peter Duerr's vision of the Human-Nature relationship", 263.

8. Gary Snyder, "The Etiquette of Freedom," in Max Oelschlaeger ed., The Wilderness Condition: Essays on Environment and Civilization (San Francisco: Sierra Club Books, 1992), 23-24.

9. "Walking," 226.

10. Henry David Thoreau, *The Journal of Henry David Thoreau*, ed. Bradford Torrey and Francis H. Allen (Salt Lake City: Gibbs M. Smith, 1984) 5:135.

11. *Walden and Other Writings* by Henry David Thoreau, ed. Joseph Wood Krutch (New York: Bantam, 1962), 172-173.

12. Journal 5:135.

13. Ibid

14. Ibid

# INDEX

## A

# B

# Z